the Flower Farmer's year

the Flower Farmer's year

How to grow cut flowers for pleasure and profit

GEORGIE NEWBERY

Published by

Green Books
An imprint of UIT Cambridge Ltd
www.greenbooks.co.uk

PO Box 145, Cambridge CB4 1GQ, England
+44 (0)1223 302 041

First published 2014

Georgie Newbery has asserted her moral rights under the
Copyright, Designs and Patents Act 1988.

Illustrations © 2014 Fabrizio Boccha

Front cover photograph by Georgie Newbery
Back cover photographs (top three) by Georgie Newbery

All interior photographs are by the author, with the exception of the following:
Pages 89 (left), 179 and 232: Nic Cretney, Maia Creative.
Page 155: Fred Cholmeley. Pages 106, 161 and 173: iStock.

Design by Jayne Jones

ISBN: 978 0 85784 233 6 (hardback)
ISBN: 978 0 85784 234 3 (ePub)
ISBN: 978 0 85784 235 0 (pdf)
Also available for Kindle

Disclaimer: the advice herein is believed to be correct at
the time of printing, but the author and publisher accept
no liability for actions inspired by this book.

10 9 8 7 6 5 4 3 2 1

Contents

Foreword by James Alexander-Sinclair 7

Introduction: Why grow cut flowers? 9

1   Getting started 15

2   Annuals 37

3   Biennials 59

4   Perennials 67

5   Bulbs & corms 83

6   Shrubs 103

7   Roses 115

8   Dahlias 127

9   Sweet peas 141

10 Herbs 157

11 Wildflowers 165

12 Cutting, conditioning & presenting cut flowers 183

13 Hedgerow Christmas 197

14 Starting a cut-flower business 205

15 Where to sell 221

16 Marketing & social media 229

Afterword 235

Appendix 1: The flower farmer's year planner 236

Appendix 2: Plant names 242

Resources 248

Index 251

To Fabrizio, the Bear and the Wriggler, who drive me to do it all.

Acknowledgements

So many people have helped us on our way that it's difficult to thank them all, but special recognition must go to Sara Venn, for strong faith in what we do, as well as for valuable horticultural advice, time and time again; also to Nic Cretney of Maia Creative, for calm perspective and strategic planning, and to Tamsin Hewer for all her Web management. Heather Edwards has taken some great photographs of what we do over the years (examples in this book!), and press driven by her photographs has very much helped along the way.

Extra special thanks go to the team at Common Farm: Sharon, Emily, Ann, Bill and Phil, who keep us weeded, pricked out, planted, picked and bunched.

Thanks also to my editor Alethea Doran, who has patiently taught me how a factual, rather than a fictional, book needs to be put together.

And to Fabrizio and our children, who put up with my absurd drive to do more in a day than most people would contemplate in a week, and with my ruthless disinterest in housework and laundry. Thanks to Fabrizio too for feeding me and making sure there is wine in the fridge when necessary.

It is our customers, especially our Twitter following, who have made this business, and I am grateful to all of them.

Foreword

Flowers are important – very important – in oh, so many ways.

Many, many years ago I gave flowers to the beautiful girl who is now my wife. Being a bit of a show-off, I felt a grand gesture was required, so delivered two hundred roses to her office in a battered Land Rover while wearing a kilt. She was extremely embarrassed, but (such is the power of young love and flowers) forgave me, and the rest is, as the cliché goes, history.

I share this snippet with you to show the persuasiveness of flowers.

Georgie Newbery is, therefore, a cross between an enchantress and Wonderwoman, as her Somerset flower farm is the source of the power to change people's lives. To make it even better, she does it with style, pizzazz and a social conscience.

Her flowers are grown with consideration for both the locality and the wider environment. Aphids are allowed (because they feed the ladybirds) and slugs are permitted (to keep the hedgehogs happy). There are hedges (for foliage and berries), trees (for catkins and bark) and a formidable miscellany of good, healthy British flowers. From sweet peas and sorrel to buttercups and bistorts: there's pretty much anything that can be moulded into a nosegay, wreath, sprig or garland.

So next time you need a bouquet (whether to get yourself into or out of trouble), eschew the garage-forecourt chrysanthemum and the scentless, pesticide-laden imported rose, and call Georgie: she can always be relied on to pack a powerful bunch.

James Alexander-Sinclair

Introduction: Why grow cut flowers?

The international cut-flower industry is a monster, a behemoth, a vast bloom-producing machine, in which plants, water, chemicals and people are bent to the will of the world's flower-greedy public. Year-round, hundreds of millions of flowers are produced by thousands of workers working flat-out to supply our desire for great big long-lasting bouquets to give our mothers, lovers, friends and neighbours.

We give flowers at the drop of a hat: for Christmas, Easter, Valentine's Day, Mother's Day; for love, congratulations, anniversaries, commiserations. . . But do we think about where these flowers come from? Do we care about the environmental impact they have had in their journey from far across the sea – through fungicide bath to air-vacuumed cold storage – and eventually, kept alive with frequent pulses of sugar and bleach, to our own kitchen tables? When a bride walks down the aisle, coyly dipping her nose to her tight-packed bunch of roses, do we think of the chemicals she's breathing in?

There is a small revolution happening worldwide. People are beginning to realize the environmental impact of their cut-flower habit. The same people who worry about where their meat comes from – how it was raised, what it grazed on; the same people who'd rather not buy out-of-season green beans or strawberries flown in from the other side of the world – those people are looking at the bunches of flowers they have, until recently, added unthinkingly to their supermarket trolleys, and they're leaving those bunches on the shelves.

They are making a quiet protest at the environmental cost of the international cut-flower trade. However, while people may no longer like to buy chemical-dunked flowers looking shocked after release from weeks of cold storage, they still like to buy flowers. And if they're not going to buy them at the supermarket checkout, or they're told at the high-street florist shop that all the flowers there are imported, then where are they going to get them?

Well, maybe you've picked up this book because you remember that your grandmothers' gardens were full of flowers for cutting. . . That there used to be a whole group of little flower farms in the next town or village. . . That small-scale, domestic flower-growing used to be a good industry. And you are thinking of turning to your seed catalogues and beginning to grow cut flowers yourself.

Home-grown cut flowers sold to a relatively local market have virtually no carbon footprint – and certainly no air miles! And however much you like to douse your garden in chemicals, believe me, you will be using an infinitesimal amount compared with the fungicide-dunking that some multinational companies impose on flowers flown in from South America and Kenya.

A mixed bouquet of flowers fresh from the garden: including roses, feverfew, ammi and scabious.

Flowers grown in a small-scale setting are fresher by days than flowers flown from big corporations' air-conditioned polytunnels in South America to the auction in Holland, then flown on again to another country's wholesale market, where they are bought by a florist who may then have them sitting around in the shop for days before selling them to you.

Grow your own and you can grow what you cannot buy: the lace-capped flowers, the wild flowers which won't travel out of water, the sweet peas which will fill your house with their sweet, peppery perfume because you don't treat them with silver nitrate to make them last longer. Your flowers will

be scented, will feed the bees, will grow in the vase. Will you be able to sell your surplus? Well, once you've read this book, then I hope you'll agree that you can.

My other half and business partner Fabrizio Boccha decided that we would be *artisan* florists and flower farmers, and I think the word 'artisan' is so important as part of the description of what we do. Our work is carefully crafted, handmade, bespoke. Our crops are grown with an eye to feeding our environment as well as feeding our own need for beautiful cut flowers. We consider ourselves artists in that each bouquet is a piece of work as perfect and fleeting as the salt mandalas made by Buddhist monks in the mountains of Nepal, whose focus on their work is intense and perfectionist, in full knowledge that the day the mandala is finished is the day that the wind begins to blow it away.

Cut flowers should have that fleeting beauty too, their inevitable end giving their presentation a loveliness more cherished because it cannot last. Artisan growers work in the moment; plan for the minute; are attuned at all times to the gorgeous combinations that can be made on a certain day, in a certain month, at a certain season – never the same twice, certainly not two years running. Let the international cut-flower trade specialize in gerberas that will last a month, in single-stem unscented roses which they dunk 50 at a time in fungicide to keep them lasting. Let them deal with the problem of extracting from the water they pour back into the system the silver nitrate in which their flowers are conditioned.

You and I can create an entirely different industry: one concerned with life, with freshness, with delicate, ephemeral beauty – safe in the knowledge that while our flowers will happily last a week in water, they won't ever last a month. Why would we want the same bouquet to last a month when we have a garden full of flowers to cut, and we need the space for a new bouquet of loveliness?

This is a book to teach you how to produce cut flowers from your patch all year round. It is also a book that will teach you the song you need to sing to sell your flowers to a public which is asking for answers, which wants to buy what you can grow in your back yard, which is already protesting against the international cut-flower trade – but which still loves a bouquet of flowers to put on the kitchen table as a weekly treat.

Who we are

Here at Common Farm in Somerset, Fabrizio and I started our little flower farm in a corner of our vegetable patch in April 2010. I have a background in writing and fashion; he in art and antiques. Neither of us are trained horticulturalists or florists – though my mother paid me to weed by the yard when I was a child, and she cut her floristry teeth working at The Dorchester in the 1960s, so perhaps what I do *is* in the genes. Half of my attraction to Fabrizio when we met was that despite lacking an actual garden, he still grew a profusion of sweet peas in pots on the little terrace outside the cottage where he lived. Together we have become excellent, self-taught gardeners, and, rather than any flower-school rules of three or five, our floristry is inspired by the Dutch seventeenth-century painters of lush still lives, and by the Scottish Colourists.

In our first season we had eight stands of sweet peas, a 3m x 3m (10' x 10') patch of dahlias and, I think, three 3m x 1m (10' x 3') beds of annuals. We grew no flowers in our polytunnel, which at that stage was still filled with a greedy crop of cucumbers, tomatoes and chillies. In our first year

we sold at farmers' markets, from a barrow out in the lane, and I had a stall at about six wedding fairs, where I learned more than I sold – about what brides want and the prices they expect to pay.

We began our cut-flower business because we needed to find a way for the house and land to pay their way. We had two very small children, and gardening is a job that can be done in 20-minute increments, which was often all the time I had. More importantly, I realized that whatever job we did to pay the bills, we were going to garden as well, and so it seemed intelligent to turn the activity that would drive us, whatever the weather, into the way we made our living.

Turning Common Farm Flowers into a serious business has been an exhausting ride, during most of which I've felt as though the learning curve has been so steep that I've been hanging on by my fingertips. I am (or perhaps was) no business-woman, and I've had no training in marketing. The many lessons we've learned are included throughout this book , so there's no need for you too to learn them the hard way!

We now sell bouquets throughout the UK for 12 months a year. We are asked for, and are delighted to supply, wedding flowers year-round. We teach something like 30 workshops a year, and the business has grown enough to support three full-time staff plus others part-time.

My point, I suppose, is that if we can do it, so can you. A patch need be no bigger than half an allotment to create income from cut flowers. After all, a sweet pea is worth a great deal more than a lettuce leaf (although, of course, it is poisonous, so perhaps its value is compromised by being *only* for show). Whether you want to grow a few bunches of sweet peas for sales from your garden gate or have your eye on creating a higher-income

A nice morning's cut from our flower garden.

business, I hope this book will inspire you, get you started, and help you throughout many years of cut-flower growing.

The human race has created a $40 billion-a-year industry in cut flowers. That's plenty of money for a lot of small growers to share!

A note to the non-UK reader

Every gardener grows plants in their own particular microclimate, influenced not only by location but by soil conditions, aspect, and so on. I have tried to be general regarding the 'hows and whys' of how you manage your plot through the seasons, but inevitably there will be moments where growers no further from us than Wiltshire may frown and mutter to themselves, "Well, not here!" So, for growers further afield, here's how I've decided to describe our year in print.

Throughout this book, I refer to the conditions needed for a given plant to grow successfully here in Somerset, and in the wider context of the UK. I imagine that the reader in the US, for example, already has sufficient interest in gardening to know that in a place such as Vermont, say, where winters come early, stay hard and thaw late, the growing season for cut flowers will be shorter than in the gentle climes just north of San Francisco.

Common Farm is in a UK zone broadly equivalent in temperature and season to zone 9, in the US system of plant hardiness zones. Most of the UK (though perhaps the far north might be considered colder) corresponds to zones 8 and 9 of the United States.

There has been much debate here about how best to describe the phases of the year so that the text is helpful to readers everywhere. The truth is that our schedule as flower farmers is quite precise, and I felt that to say that we did anything in 'mid spring', for example, was too vague – when we schedule our getting-out the dahlia tubers from their winter sleep for 1 April exactly. I hope that readers elsewhere will forgive this decision, and I am sure that, wherever you are, you are intelligent enough to extrapolate your own gardening approach from my account.

Appendix 1 includes a year planner for the flower farmer, and in the table there each month is also given in a 'generic' form, to help you translate the timings described in this book to suit the seasons in your own location.

Horticultural Latin

I have aimed to refer to plants by the name they're most commonly known by – whether that be Latin, anglicized Latin, or their common name. Of course, where common names are used, the plant might be known differently in other places, so please see Appendix 2 to identify a plant by its Latin name if necessary.

And finally. . .

The lists of plants I provide in this book are by no means exhaustive. Who could possibly list all the annuals one could grow in a year, let alone all the perennials and shrubs in the world which make good cut-flower material? I could spend years adding to the lists in this book. Moreover, I feel strongly that the way you curate the collection in your cut-flower patch is up to you. My taste in roses is almost certainly going to be quite different from yours; likewise my taste in tulips, viburnums, foxgloves and the rest.

My aim in this book is rather to inspire you to look beyond the obvious; to ask questions about what will and won't make good floristry material; to turn your ability as a gardener into an artistry with flowers. Whether your desire is to have a cut-flower patch that you harvest on a Friday and sell at a regular Saturday market, to sell bunches of sweet peas at the garden gate, to supply wedding flowers, or to take on the big flower importers with your locally grown blooms, this book is intended to inject a can-do attitude into your gardening.

You will learn as you grow. Make notes – and remember that no two gardens are the same, and no two horticulturalists will tell you how to grow the same way. Gardening successfully, and therefore flower-farming successfully, is more about being in tune with your plot than anything else. Listen to what I have to say, listen to all the other flower farmers – then do what you want to do and make your farmed garden work for you.

There's no reason why you, like me, shouldn't farm your garden, make the world a more beautiful place, and feed the bees while you do it. Good luck and good gardening!

Chapter one
Getting started

The blank canvas – the freshly fenced-off acre of a field – may be an exciting start to building your dream, but can be rather daunting in its emptiness. The trick is to chop your space into carefully planned, manageable chunks, and all of a sudden you'll have a strategy. Plant your ideas one small seed at a time, and watch them grow and bloom.

The first cutting patch at Common Farm, which was cleared by pigs. Note the soon-to-be wind-protecting curve of native hedging settling in among the buttercups.

You need surprisingly little space to create a really productive cut-flower patch. At a lesson I taught in Northamptonshire recently, we assessed an ordinary-sized domestic garden space and identified a patch of the vegetable garden that would happily turn itself over to flower production.

In a bed of about 3m x 3m (10' x 10'), we saw room for perhaps 15 sweet pea plants, five ammi, five cosmos, five dahlias, five sunflowers, a row of cornflowers, a row of pot marigolds and a row of hare's ear. There were already roses in a herbaceous border edging the vegetable patch, and the herbs growing at the other end of the garden would make lovely scented foliage for the grower's posies. Among the group I was teaching were two garden designers, a florist, a baker, a writer, a lady with an allotment who wanted cut flowers for her house, and another who wanted to sell mixed bunches from her front gate. None of these people would call themselves 'eco-warriors' – they were simply sensible people who had looked at the international cut-flower trade and didn't like what they saw. They realized that if one is a gardener of any kind, one can grow cut flowers – which will not only satisfy the aesthetic desire for cut flowers in the house, but also feed the bees.

So, you have your dream – or even your patch, your meadow, your corner of a rented field. It's time to lay out your plot. Go and stand in it, sniff the wind, test the earth, and think hard before you apply the sharp edge of spade to soil. The better planned your plot is, the harder it will work for you.

Plot design and practicalities

Whether you're planning a single bed for small-scale growing or several acres for commercial production, the same design rules apply. Think carefully about how you will lay out your patch, and you will save yourself time, energy and (most of all) irritation later on. A carefully planned garden will reward you time and again for the work you put into getting the mechanics of it right.

How much space?

If you're growing to sell at your gate, at farmers' markets or to local florists, beds of 3m x 1m (10' x 3') will never be too long to walk around or too wide to reach across without treading on. It's a good idea to begin with one bed cut into three (or three whole beds, and so on), since if your beds are divisible by three, you're set up for a plant rotation system (see page 21). If you then lay out another bed to the windward side of your first bed, to help with wind protection (see page 19), you are ready to plant a substantial crop in three annual beds and one perennial – all possible in less space than the size of an average allotment.

Our first flower patch – about the size of three allotments – provided enough for us to run plenty of market stalls and do five weddings in that first year.

We had a 6m (20') polytunnel, an effective wind-proofing hedge, and about twelve 3m x 1m (10' x 3') beds.

Cutting your space up into small sections also makes the job less daunting for any grower, as you can achieve great things by dealing with one small section at a time. A great ploughed plain before you is daunting, whereas a series of small areas that can be dealt with one at a time is a list of surmountable challenges.

A fan of mulched raised beds ready for planting in spring. While an attractive design, the beds are unequal in length and difficult to get a mower in between – you could be more organized!

What to grow

There's no need to try to include a wide variety of flowers in your patch. You can be a professional cut-flower farmer and grow as few as ten different varieties – the skill is in being clever about which ones you choose. The advice and recommendations in this book should help with making your choice, but remember that the most important thing is to grow flowers that you love. A patch of something you grow because you think you ought to, but which you don't love, will be ignored and will cost you money in wasted space. I never grow cleome because I hate the smell and the spines. I could think that I ought to include it, because it's a classic annual to grow for cut-flower production, but I've learned – to my cost in space and time – that I won't pay it any attention other than to be annoyed by it.

Last year in our garden we planted up two 3m x 1m (10' x 3') beds as follows:

- Bed 1: Lots of *Ammi majus* (we use a great deal of ammi in our summer flowers) and three rows of cerinthe (a generous cut flower, which we use largely for its glaucous blue-green leaves, but the stems can get straggly so we plant successively through the summer and put the plants out in small chunks as space fillers).
- Bed 2: A mix of pale blue cornflowers, larkspur, orlaya and annual scabious.

These beds started flowering at the end of May and we cut from them until September – you can expect to get a maximum of two months' useful production out of an annual plant.

Both the beds have low hedges of young box plants at one end, and one of them has a small hedge of *Spiraea japonica* 'Candlelight' at the other end.

So from just this small planting area we had cut flowers and greenery to take us all through the season. I could have planted fewer ammi and put a stand of sweet peas at one end of that bed instead, and then this really would have been a great cutting mix for posies, for a farmers' market, to supply buckets of flowers for do-it-yourself weddings, and so on. . .

Windbreaks

Cut flowers are, of course, valued for their looks – which are easily ruined by the weather. And so protection from the weather is what they need most. Before laying out your cut-flower patch, first consider the wind: where it comes from most of the time, and how you can stop it flattening your cosmos, dahlias, sweet peas and so on.

Don't plan to build an expensive solid wall as a windbreak. The wind will rush at it, hit it, be forced up over it and then crash down, flattening whatever's on the other side. People with walled gardens often complain about odd areas where eddies of air become whirlwinds in bad weather, causing all sorts of damage to their precious plants. It's best to construct a windbreak that allows some air to pass through.

So think of using hedging, or trellis fencing, or even, in the short term (while your hedge grows),

June in the patch. Tall native hedging makes an effective windbreak on a more blustery day.

some green mesh netting for windbreaks. Or – since few people have more gardening space than they know what to do with – it can be a good idea to create your windbreak with plants that you can later use as foliage to accompany your flowers.

A patch of something you grow because you think you ought to, but which you don't love, will be ignored and will cost you money in wasted space. I never grow cleome because I hate the smell and the spines. I could think that I ought to include it, because it's a classic annual to grow for cut-flower production, but I've learned – to my cost in space and time – that I won't pay it any attention other than to be annoyed by it.

Useful hedging shrubs for cut foliage and flowers

❋ Both the acid yellow and the black varieties of physocarpus.

❋ Choisya (though some people hate the smell).

❋ Some viburnums are useful as hedging, as well as for cutting – especially evergreen and winter-flowering varieties, though I love the wild guelder rose and the snowball bush varieties too.

❋ Pittosporum – variegated and black varieties are stunning, though they can be frost-tender.

❋ Euonymus – both the yellow and the white variegated ones are very useful at Christmas for wreaths, though euonymus are slow-growing and not good if you need a tall hedge.

❋ Brachyglottis – useful all year round. Cut off the yellow flowers if you don't like them.

❋ A tall rosemary like 'Miss Jessopp's Upright'.

❋ Lavender for low hedging.

Buying tip: Invest in a few good-sized plants to start off with, then take cuttings to increase your stock. See Chapter 6 for more on shrubs to grow for cut foliage and flowers.

A low hedge of physocarpus provides a mid-plot windbreak between sweet peas and ammi.

One way of doing this is to dedicate some beds down the windward side of your plot to perennials. This creates extra wind protection for your annuals and biennials, helping them establish without suffering 'wind rock' (damage to roots caused by movement of the stem in the wind). If you don't have space for this, perhaps establish some bushy perennials or shrubs at the windward end of your beds, for both greenery and wind protection.

Rabbits and deer

If there are rabbits or deer in your area, you'll need to fence them out. Deer love rose tips and new growth on shrubs; rabbits will mow your crops as if they were a lawn, and ring your trees and shrubs effectively enough to kill them. Each sweet pea *flower* is potentially worth as much as 65p to you, so don't let the rabbits have them! Edge your plot with chicken wire, buried underground horizontally for 30cm (12") under about 2cm (approx. 1") of turf, and you should be able to keep rabbits out (see the illustration opposite for finer design details). Don't forget to line your gates with chicken wire too, or the rabbits will squeeze under or through the bars. When it comes to deer, fencing will need to be 1.2-1.5m (4-5') high to deter them.

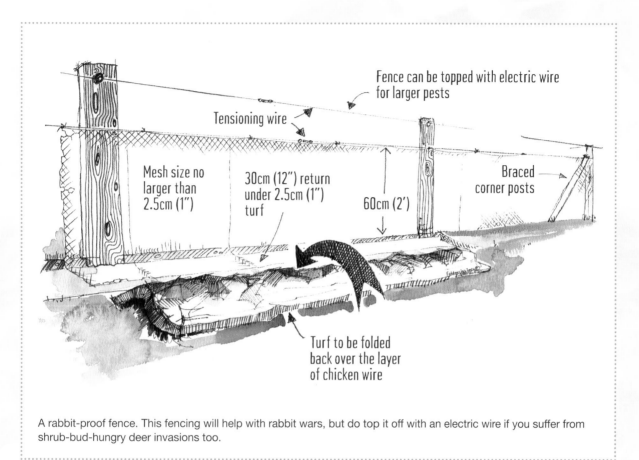

Fence can be topped with electric wire for larger pests

Tensioning wire

Mesh size no larger than 2.5cm (1")

30cm (12") return under 2.5cm (1") turf

60cm (2')

Braced corner posts

Turf to be folded back over the layer of chicken wire

A rabbit-proof fence. This fencing will help with rabbit wars, but do top it off with an electric wire if you suffer from shrub-bud-hungry deer invasions too.

While deer fencing is costly, shrubs and roses are an expensive investment that you would probably rather not lose to nibbling night-time visitors.

I have heard that planting autumn-flowering crocus around your cut-flower garden will repel deer and rabbits, and it does sound like an attractive solution. I have a friend who stands a length of fluorescent strip lighting in the corner of her cut-flower field, attached to a solar-powered battery to flicker and put off marauding varmints – again, I've never tried this, and if you have neighbours overlooking your patch, it might be irritating to them. Alternatively, dousing around your plants with chilli and garlic dip (see Chapter 5, page 86) will discourage rabbits and deer. Personally, I prefer just plain fencing: it is an expensive investment, but time spent making chilli and garlic dip for protecting plants individually might drive you madder in the end!

Planting rotation

As with all crops, it's a good idea to design your plot with a rotation in mind, for the same reason as you would cabbages, peas and potatoes around vegetable beds: to avoid the build-up of pests and diseases. If you arrange your beds in threes, with a fourth bed on the windward side for perennials, then that's a pattern you can work with as you expand your patch, or can scale up if you're creating a bigger patch in the first place.

The flowers you will be growing for cutting won't fall neatly into 'family groups' in the same way that veg do, so the way you arrange your rotation won't be as categorized as it might be for veg: the idea is just to avoid growing the same species on the same piece of land from one year to the next. You can choose the plants you're likely to grow most of and base your rotation on them: for example, a rotation of sweet peas, dahlias, and a bed of other annuals is a good mix; or a bed of biennials, a bed of sweet peas, and a bed of some other annuals; and you could add a perennial bed of roses under-planted with herbs and alchemilla, for example.

Since sweet peas are legumes, as are any pea or bean, they will fix nitrogen in your soil, so if you're going to grow them, use them as number one in your rotation plan. Your other annuals, your dahlias or your biennials can then follow your sweet peas in rotation, year-on-year.

Planting into high-sided raised beds protects seedlings from wind rock.

Raised beds

You should consider 'flower farming' your patch in raised beds for a great many reasons, not least because in your first season or two the scale of your task may seem daunting, and, as noted earlier, a daunting job chopped into manageable sections becomes possible rather than over-whelming. You can weed a small raised bed in 20 minutes. You can plant a small raised bed in 20 minutes. You can cut all the flowers available from a small raised bed in 20 minutes. If we'd grown flowers in larger areas when we started, I know I would not have been able to see the individual jobs to do because of the overwhelming scale of the area in which they needed to be done. There are other advantages of raised beds too:

* If you lay out three, six or twelve beds at a time, you've always got a three-way rotation,

which makes it easy to manage your yearly planting plan.
* You can optimize the size of your beds to avoid ever needing to tread on the soil.
* If you're gardening on clay, you can dramatically increase drainage and improve the soil condition by 'making' the soil yourself – with the addition of well-rotted horse manure and garden compost – within the confines of a raised bed.
* Ditto for chalk or gravelly soil: you can increase fertility and water retention by feeding them with a great deal of well-rotted horse manure and garden compost.
* The beds offer some protection against wind rock for newly planted-out plants, and make it easier for you to give some protection to seed-lings, with fleece or shade mesh.

How to make raised beds

❋ Lay out 'boxes' of preserved wood, of 3m (10')-long sides and 1m (3')-long ends, and nail them together.

❋ Roughly rotavate the space inside the boxes. If you're gardening on clay, sprinkle with a little lime to help break up the clay.

❋ To stop perennial weed roots from sprouting through your new earth, lay several layers of cardboard (stripped of all parcel tape) on top of the rotavated earth.

❋ Fill your boxes with topsoil or municipal green-waste compost (see page 26). The latter can be inconsistent in quality, but it's usually good enough to use to fill a raised bed, it is peat-free, and it won't be full of weed seeds, which topsoil may be.

❋ Add a layer of well-rotted garden compost or well-rotted manure to the top of the bed. This will serve as a mulch, provide insulation and help with water retention, and, once the worms have dug it in, it will improve the soil. Keep the soil in your raised beds in good 'heart' by mulching with well-rotted horse manure at the end of the season when you've cleared away your annuals, by mulching with comfrey leaves when you have them to spare, and by feeding with compost tea throughout the season (see page 28).

❋ Keep the surface of the soil in your raised beds 10-12cm (4-5") lower than the sides of the beds, so that the beds themselves provide a little protection from wind rock when you're planting out, and so that you can cover the beds with fleece for frost protection or mesh for sun shade and still have space for newly planted-out seedlings to stand proud of the soil. Late-flowering tender annuals in particular seem to hate being planted out, and will sulk or fail unless given a bit of cosseting.

Staking

There are three basic options for staking plants, whether annual or perennial.

Pea netting

Provide horizontal support with pea netting laid flat at a height of about 1m (3'), tied to stakes at regular intervals along the bed (as pictured overleaf). This is an extremely efficient way to support your crops and is not difficult to put up. You might think that this kind of netting might get in the way, but you can hoe around the plants easily enough under it, and when cutting your flowers, the spacing in the netting is wide enough for you to easily reach down through it to get the maximum stem length. It is more annoying to lose a whole crop to a big storm than to have to work around a little support netting through the season.

String or twine

The greener version of pea netting is to stake the edges of your beds at regular intervals and run a zigzag of garden twine horizontally along the bed. It won't be as strong as the pea netting, but it works well on a small cut-flower patch, and is an acceptable solution if you are very conscious of the ecological 'footprint' of your gardening.

Delphiniums growing through horizontal pea netting for sturdy, functional support.

Stake plants individually

In my view this is the least effective approach: it is a fiddly waste of time. You can make teepees of bamboo or hazel clippings for roses to grow through, and a similar arrangement will work well if you have, say, four or five dahlia plants. But for a whole bed of dahlias, staking with one of the netting arrangements described above will give your plants much greater support and save you a great deal of time. If they have grown through netting or a zigzag of string they'll survive a storm much better than if they can break their individual supports, or pull them up. A serious August storm will flatten a whole dahlia bed if it's not properly supported, and this will hurt much more than the work that goes in to providing them with a proper structure to grow through earlier in the season.

Soil, compost and feeding

Without soil in good heart (and a pretty neutral pH of about 7), your success as a grower will be limited. Cut-flower growing is a greedy business – especially in the case of annuals, which need a lot of nutritional support in their race to germinate, grow, flower and go to seed in one season. Remember that however you choose to make your beds, you *are* creating soil structure. You will inevitably add manure, compost, leaf mould, etc. – and that's before you get to compost tea and seaweed solution, lime, etc. . . . And what will you mulch with? 'Strulch' (straw-based compost – good for keeping plants warm in the winter), municipal green-waste compost, spent mushroom compost. . .?

So, you will be creating soil. As the seasons turn, and perhaps twice or even three times a year, your beds will be cleared, replanted and expected to perform again. Every time you do this you disturb the soil structure, and create a new structure for your next crop to grow in.

I recommend Charles Dowding's books for advice on no-dig gardening and building no-dig raised beds. You could also look at the Ten Minute Gardener on YouTube for a demonstration of building no-dig beds. That said, we don't do 'no dig' here at Common Farm. This is partly because we garden in a meadow where creeping buttercup, pendulous sedge and couch grass invade our beds from every side throughout every season, and Fabrizio and I are both obsessive about getting every last bit of root out. But it is also, mostly, because our long-ingrained habits involve a border fork (me) and a rotavator (him). At the end of the day, you are making a cutting patch which must first and foremost suit *you*. Going too hard against your own gardening habits may actually slow you down. So, we rotavate our beds hard when we create them (we have to mix in a great deal of drainage material and compost to break up our solid clay soil), but once the beds are made we don't dig them hard again at all. We simply mulch generously once a year, and feed the soil and plants often.

In a small patch, so not too great an area to manage, experimenting with no-dig is a good idea. Perhaps, if you start with three rotating beds, you might have one no-dig bed, one half-dig (which I suppose is what we do – lots of layering, but never double-digging) and one double-dig (as your grandfather might have taught you).

A note on the law

If you plan on selling plants, then beware compost mixed with animal manure of any kind. In the UK DEFRA has strict rules relating to potential pathogens spread in manure (and I imagine food and farming authorities around the world have similar rules). So plants for sale need to be sold in recycled plant-matter compost.

New plant cultivars may be subject to 'PBR' (plant breeders' rights), which is like copyright for plant breeders and gives them (international) financial rights in the form of royalties. This applies to selling not only propagated plants but also harvested material – cut flowers and foliage. Usually plants are labelled if they are subject to PBR.

Top tips for good soil management

Follow these basic principles for successful growing:

❁ **Mulch.** At least once a year, give your beds a good layer of well-rotted manure and/or garden compost.

❁ **Test.** At least once a year, test your soil pH. This simple precaution will help prevent you scratching your head and wondering why your flowers aren't successful that year. A balanced soil pH will give your plants the best medium to grow in.

❁ **Feed.** Regularly feed your *soil* with compost tea. A small cut-flower patch can be fed with a washing-up bowl full of compost tea once a fortnight in season. Compost tea feeds the soil with good bacteria and encourages the growth of mycorrhizae: a happy symbiosis between plant roots and soil. Plants grown in fed soil grow faster, bushier and stronger, and are better able to fight everyday battles against the likes of greenfly, slugs and mildew.

Soil pH

It is always worth testing your soil pH: testing kits are cheap to buy from any garden centre. Test the soil in several different areas of your patch – a limey soil in one corner does not promise limey soil elsewhere. And don't assume that once you've tested your soil and done something to balance its pH, if it's very acid or alkaline, that the pH will stay the same from one season to another. The plants you grow, what you feed them with, and the compost and manure you add will all have an effect on your soil balance. So check the soil pH once a year or so.

Be aware that spent mushroom compost and municipal green-waste compost can be alkaline. Even so, at Common Farm we do use municipal green-waste compost, because it's easy to source locally and because it's peat-free. It is useful for building up beds and as a mulch – yes, it's fibrous and not necessarily very nutrient-rich, but nutrients are added in garden compost, manure and liquid feeds.

Call your local recycling centre to find out where you can get green-waste compost in quantity. It will cost almost as much as the compost again to have it delivered, but if you can hire a trailer you should be able to collect it from your nearest depot and pay for it according to its weight. It's worth testing the pH, either when you go to buy some or when it's delivered.

If you're buying in topsoil to fill raised beds, first ask about its pH balance, as well as its sterility (re weed seeds), where it's come from and what it's been used for. The more questions you ask, the more confident you can be that any soil you're bringing in to your plot is of a high standard. Again, check its pH when it arrives.

Garden compost

Keep a space for a compost heap in your cut-flower patch. A compost heap is so much more than just a pile of horticultural debris: it is the source of a great deal of food for your garden. A well-managed compost heap can make you new, nutritious soil in as little as six weeks, though we tend to wait for about a year for ours.

It will also be the home of a great many good predators of pests: toads may shelter in its cool,

Making a compost heap

❀ Make three bays out of something as simple and easy to get hold of as old pallets.

❀ Fill your first bay with a combination of grass cuttings (a good supply of nitrogen – 'green' material), dry garden clippings, cardboard, paper, straw (a good supply of carbon – 'brown' material), kitchen waste and weeds. Include coffee grounds and egg shells in your mix, as slugs don't like them (you can of course use these directly around plants, but it's also good to turn your compost into an anti-slug world).

❀ Be sure to bake your eggshells before adding them to the compost heap: rats and mice love the nutritious, high-protein membrane which usually stays stuck to the shell when you crack it – bake your eggshells and you cook off this membrane, making your heap less attractive to rats. Baking egg shells also makes them harder and crunchier and even less pleasant for the evil slug to cross when seeking out your young delphiniums.

❀ Make your compost heap in layers and when the first bay is full, turn it, cover it in a layer of hot grass cuttings and move on to the second bay. By the time the third bay is full, your first bay should be ready for use.

damp edges; hedgehogs may nest around the back of it. Deep in the warm centre of a compost heap you may sometimes find a slow-worm or a grass snake. A compost heap is larder, shelter and friend to everything from that hedgehog, gorging himself on your slug population, to invertebrates and good bacteria which feed your soil. It is as necessary a part of a productive garden as are the plants. Shop-bought compost can be useful if you want to raise seeds in a sterile medium, but otherwise the home-made variety will be a great deal more nutritious for your earth and therefore your plants.

Including a compost heap structure in your plot will also give you another little edge of wind-proofing. Keep space behind your compost bins for a few comfrey plants, with which you can make a high-potash tea to feed your flowers later in the summer (see overleaf).

The compost itself can also be used to make compost tea, which you can use to feed your soil with not only nutrients but good bacteria – making it a happy home for your plants when you're ready to plant them out.

Home-made liquid feeds

Another important ingredient in your cut-flower plot is a small clump of nettles – make sure you have some behind your compost heap or in a similar tucked-away spot. These will feed the butterflies, and will also give you the ingredients for nettle tea – a high-nitrogen feed to get your plants going early in the season.

Feeding your earth with compost tea, and your plants with nettle tea, will help them fight off slug and other potential pest damage as they grow. Feeding with comfrey tea after midsummer will encourage strong flower production.

Compost tea recipe

In a washing-up bowl, put:

- ❋ a handful of fresh cow, horse or chicken manure
- ❋ a handful of home-made garden compost
- ❋ a handful of nettle tips.

Fill the bowl with water and leave it somewhere warm where you will pass relatively often.

Stir the contents of the bowl every few hours, looking for yellowish scummy bubbles rising to the surface (these show that the bacteria are working to make your tea nutritious).

In 48 hours your tea should be ready for use (in cold weather this may take a day or two longer). Dilute it 1 part compost tea to 10 parts water, and water your beds with it.

Make this tea year-round for a constant earth feed to encourage good growth.

You can tell when compost tea is 'making' by the scum of yellowish bubbles popping on the surface.

Nettle tea recipe

Fill a watertight bin (an old kitchen bin will do well) with new-season nettles and then pour water over them until the bin is full.

Cover and leave to steep for three weeks, or until the stink is almost unbearable.

Dilute it 1 part nettle mix to 10 parts water and feed your plants with it. This highly nutritious nitrogenous feed will promote green growth and is especially suitable for use early in the season, when you are establishing new plants in the garden.

Make this tea in late April / early May and use until midsummer.

Comfrey tea recipe

Fill a watertight bin (an old kitchen bin will do) with roughly chopped comfrey leaves and stalks. Cover with water and leave to steep for three weeks or until the smell is almost unbearable.

Dilute it 1 part comfrey mix to 10 parts water and feed your plants with it. This potassium-rich feed will help your garden plants flower and flower, right through until the end of the season.

You can also use chopped comfrey as a good midsummer mulch which will rot quickly into the soil. Slugs don't like it.

Make this tea mid-June and use until the end of the season.

Propagation

You may plan on being a ruthless direct-sower of seed, but I do recommend that you propagate seed in trays too: to beat the slug, the rain, the frost, the wind – tray-sown plants under cover will not only be your backup but will also extend your season, giving you bigger plants to put out in the field as the soil warms up. Success in this business relies a great deal on you hedging your bets. And with propagating in trays you can successfully plant small quantities of the same annual, for instance, or manage the inevitable hundreds of minute nicotiana seedlings which will prefer pricking out and protecting until after the last frosts are past to being direct-sown in the ground.

A heated propagating bed, like the one shown in the illustration below (see overleaf for more details), is an inexpensive way to get ahead with your propagation and to keep your sprouting seedlings frost-free. I remember researching heated propagating mats to buy, and they were incredibly expensive for very small areas (around £70-£80) – much too much for us at the time.

We plant in fortnightly chunks throughout the season, and from January to the end of March our propagating bed is kept at a warm temperature. We have propagating areas outside too at Common Farm, but we wouldn't be without this warm bed in the greenhouse to encourage quick germination in our sown seed trays.

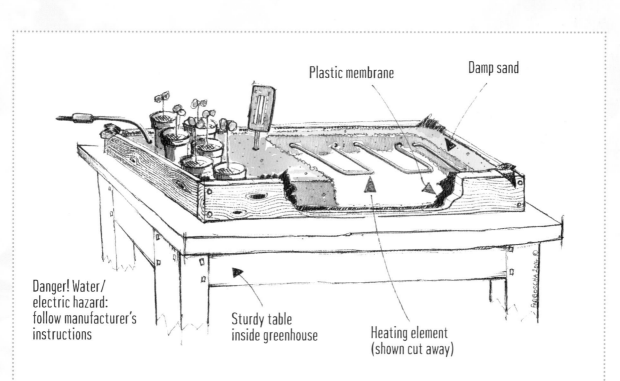

Plastic membrane

Damp sand

Danger! Water/ electric hazard: follow manufacturer's instructions

Sturdy table inside greenhouse

Heating element (shown cut away)

Fabrizio's home-made heated propagating bench. The bench illustrated here cost us the price of the wood, the sand and the heated cable – I think the heated cable was £8.

How to make a sand-box heated propagating bed

❋ Using lengths of 'two by four' (2" x 4") timber, make a rectangular box on a sturdy frame to fit the space you have. Line the box with thick polythene, up to the top of the timber. Half-fill the box with sand and tamp it down until it's flat.

❋ Lay a heated cable in a gentle wiggle across the box. These heated cables are easy to buy online. Do follow the manufacturer's instructions when plugging your cable in.

❋ Fill the rest of the box with sand and tamp it down. Water the sand and switch on the cable. It'll take a couple of hours to warm the sand around it – you'll feel the difference relatively quickly. Put trays of freshly planted seed to sprout on top of the newly warm bed.

❋ Never let a sand-box heated propagating bed dry out when seeds are germinating on it: sand gets very hot when it's dry, and you don't want to burn the toes of precious seedlings.

Crop protection

You need to think about how you're going to protect your crops once they are ready for planting out but not robust enough to be exposed to the weather. Young plantlets are fragile and easily bashed by wind and rain.

We inherited a small domestic greenhouse, so one of the first things we did was make a protected cold frame beside it, for hardening off seedlings and dahlias before planting them out. Now we have two greenhouses and they are arranged in such a way as to give further protection to the original cold frame, and as our business grows we keep adding more cold frame areas, which are used all year round.

Our cold frames are constantly fed by material coming out of our greenhouses and polytunnels, both of which are often at full stretch, especially in the spring.

A polytunnel or a greenhouse?

If you are planning to produce cut flowers on any scale and, perhaps more importantly, without risk of losing crops to the vagaries of the weather, you will also need some sort of covered growing area – that is, a polytunnel or greenhouse.

The main pros and cons of these two options are summarized on the following pages, but the obvious difference is cost: you can usually get more polytunnel than greenhouse for the same money – and so if you must choose one over the other, the tunnel is the better investment. However, a greenhouse is so much more attractive that, in the unlikely event that I will ever be able to afford it, I will replace my tunnels with glasshouses. It is worth keeping an eye on Freecycle or the gardening pages of your local advertiser: greenhouses do come up 'free to a good home', and so it is possible to install them inexpensively.

A greenhouse is much better as a propagating environment, as it is easier to control temperature

If at all possible, you could have a small greenhouse for propagating and for plants which like it hot, and a bigger polytunnel for growing on plants and cropping. You can then propagate in your greenhouse on inexpensively made propagating beds before moving propagated seedlings or young plants into your cooler tunnel.

and moisture levels, simply because you can open and close windows, and also because it is usually smaller. If at all possible, you could have a small greenhouse for propagating and for plants which like it hot, and a bigger tunnel for growing on plants and cropping. You can then propagate in your greenhouse on inexpensively made propagating beds (see box opposite), and move propagated seedlings/plants into your cooler tunnel.

Large-sized polytunnels may need planning permission, so do contact your local authority before ordering one. Even if they're classed as 'temporary' structures, tunnels can be hard work to put up and to take down again: moving a tunnel when it's been erected will be a great waste of your time.

Polytunnels

Pros:
* More space for your money.
* Can extend the growing season for up to a month each end of the summer.

Cons:
* You need to replace the poly covering every ten years.
* They can blow away, so need to be very strongly erected.
* They can't be used for built-in potting benches, etc. that need solid walls for support.

* You may need planning permission.
* Second-hand tunnels can be more trouble than they're worth: hoops may have been pushed off kilter by being moved, and their feet often arrive encased in old concrete clumps. It's often easier to buy new.

Our polytunnel in full production in June.

Greenhouses

Pros:
- Good for a controlled propagating environment: easy to control temperature and moisture levels.
- Good for raised potting-bench areas, for working at waist-height or table-height with storage space underneath.
- Second-hand greenhouses are easy to find and not too difficult to erect. If a couple of panes of glass break in the moving of a second-hand greenhouse, greenhouse-sized panes are usually kept in stock by local glass suppliers. Be sure to ask for safety glass.

Cons:
- Less space for your money (roughly less than a quarter of the space for the same cost).
- For this reason, often too small to be growing full crops in.
- Vulnerable to damage, as glass is fragile.

Siting your structure

Growing cut flowers, as with any kind of gardening, is to a great extent driven by priorities, time consideration and finance. But where indoor growing space is concerned, if you can design your arrangement of greenhouses or tunnels with the potential for versatility and expansion, however small you start, you will be grateful for it as the years go by.

You can think of your tunnel or greenhouse as a way to make yourself a protected outside corner: a spot useful for cold frames, for a potting bench, and for plants that you may not want to put anywhere more exposed until the last frosts are over (see illustration on page 35). Make use of your structure as additional wind protection for outdoor crops, siting it windward of your beds. The swoop of the gently curving roof of a polytunnel will help dissipate the wind and stop it crashing down on your flowers.

Remember that both polytunnels and greenhouses need to be put up with great care, because the minute there's a line not straight or poly not pulled taught, there's a weakness which the wind might find. So, for both types of structure, over-engineer their attachment to their base, and in a bad storm cross your fingers that you've done enough.

Whichever structure you choose, buy the best and biggest you can afford. The greenhouse we inherited, 2m x 3m (6'6" x 10'), is now fitted out with two cheap heated propagating beds. We then built a 5m x 6m (16' x 20') polytunnel. This has now doubled in size, expanded over the beds we'd built to fit behind it. We were then offered a 3m x 4m (10' x 13') greenhouse by a neighbour, and these three structures were arranged to make a sheltered area in which we had our cold frame, which was a simple L-shaped arrangement of high-sided boxes, lined with black mulching fabric. We've now built a new potting bench and cold frame along the side of the first tunnel and put up a new 6m x 20m (20' x 66') tunnel alongside it.

How to use your polytunnel

I focus on polytunnels rather than greenhouses here because a) if you do go to the trouble and expense of putting up a tunnel, it will take up a considerable amount of space, which you won't want to waste, and b) I hope you've gone for a tunnel rather than a greenhouse. For your money you'll always get a great deal more tunnel than greenhouse, and while I like my greenhouses, and am grateful for the extra space they give me, I would be able to run my business without them. I would struggle without my polytunnels.

In essence, a polytunnel is for:
- propagating plants
- growing crops
- hedging your bets against the weather.

Our small polytunnel in April – not an inch of space left unused.

When planning your tunnel space, it's important to keep the space versatile. So, make any potting benches you have in your tunnel easily removable so that you can use their space for growing during summer months. Make raised beds in your tunnel in such a way that you can put up trestles or hang potting benches from the bars above the beds and then take them down again. During the winter you can use the space under benches for forcing: if you're growing hyacinths for cutting, they will grow taller in search of light (and therefore be more useful as cut flowers) if they're grown shaded under a potting bench.

Design your tunnel with rotation in mind, as you would your outdoor beds. You could make beds on three sides of a tunnel and, if it's wide enough, add another down the middle.

Ventilation

The best kind of tunnel will have a great amount of air moving about. If you order a tunnel with green mesh sides up to about 1m (3') high, then you can have poly screens to roll up and down over the green mesh, depending on the weather. This system is well worth the expense. A cold tunnel full of flourishing seedlings will be warm in March with the sides rolled up during the day, and if a sudden cold snap is forecast you'll be able to protect your plants by rolling down your poly sides

Anemones flowering in March in a cold tunnel with mesh sides and no poly to roll down.

and using some heating inside on cold nights (a gas burner or electric heater; even just tea lights in jam jars dotted about the ground will keep a light frost at bay). You don't want any build-up of disease or mould spores in your tunnel, and a well-aerated space will really help with this.

Heating

Run warming cable through the soil in your beds, which can then be turned on to make your tunnel into a giant heated propagating bed. This is also a great way to stop the frost landing inside the tunnel: heated cable just keeps the edge of the cold off. Do be careful, though, to keep the earth well watered: dry earth heated by cables might burn the roots of your precious seedlings, which will already be struggling if the earth around them is too dry.

Water

Add a watering system. This will save a great deal of time. If you rig up guttering along the sides of your tunnel and put water butts to catch the run-off, you can then make a drip system leading from these butts to run along the earth surface of the beds. Be careful to block the end of your pipe, so that the water doesn't just run through and flood one area of your tunnel's beds. And ensure that the other end of this pipe is attached to a tap that can be turned off. Watering this way is better done when the butt is full, as the weight of the water will push it all the way round your raised bed systems. If you let the water drip through an open tap as and when the rain comes, you might find that all the earth at the butt end of the pipe is soaked and the rest of your tunnel beds don't get watered at all.

Soil in polytunnels and greenhouses, especially if you make raised beds, can get incredibly dry, so be prepared to spend a little time literally standing there with a hose if you can't afford a watering system in the first instance (our business is four years old and we've only now invested in proper irrigation). Keep the soil in your tunnel well mulched to prevent water evaporating from the surface: a dry tunnel environment can leave your plants desiccated in no time, and while the surface may look wet, it's worth just digging into it with a trowel and checking moisture levels 15-20cm (6-8") down. Once a year, a good surface mulch with well-rotted organic matter, whether it be manure or compost, will feed your tunnel beds and help them to retain water.

Pests and disease in the tunnel and greenhouse

Cleanliness really is next to godliness in a covered growing environment. Outdoors you have the wind, rain, strong sun and lots of predators to blow away, drown, burn off and eat pests or combat disease. In an enclosed environment, however, you do risk the build-up of all sorts of trouble. Don't lose heart, though – with a little regular housekeeping all will be well:

* Wash your pots with a drop of bleach or a drop of white vinegar in the water to kill off mould. Re-using seed trays and pots is eminently sensible, but they aren't dishwasher-proof, so don't try that. Washed and dried pots can be stacked somewhere in a corner out of the way ready for use again.
* Wash the inside of your polytunnel or greenhouse once a year, again with a drop of bleach or white vinegar in the water. This will get rid of the algae that will be greening the surface and so ensure that the sun can get through to promote growth, as well as kill off any mould that may be growing. Mould spores may encourage botrytis and damping off, and you want your precious seedlings to live, not shrivel

up and die on the surface of seed compost you've so carefully made for them.

- Keep an eye out for greenfly and other aphid pests. If you catch an infestation early, you can just squidge them between your fingers. And you can introduce ladybirds, who will love them (see Chapter 11, page 168).
- Leave the doors of your tunnel or greenhouse open on warm days to encourage air movement and to welcome flying predators.

Mice

Lastly, a few words about mice, as they can be the bane of your life in a polytunnel or greenhouse. A tunnel is a warm, relatively dry environment with easy soil to burrow into, and full of seeds to eat – mouse heaven! A germinating sweet-pea seed is the most delicious thing on Earth to a mouse, quite closely followed by sprouting sunflower seeds, and so on. A mouse will happily dig through all your seed trays in search of its favourites, leaving the less popular seeds dug up and drying out on the surface. Not all of us have a grass snake kindly moved in to our tunnels, so usually the only option is to put out mousetraps.

Early in the season, when you have trays of seedlings germinating, you might consider suspending your seed benches from the bars reinforcing the top of your tunnel, so that the mice have no table legs to climb. Or you could cut plastic cups in half and tape them around the legs of your potting bench, so that the mice can't get past them.

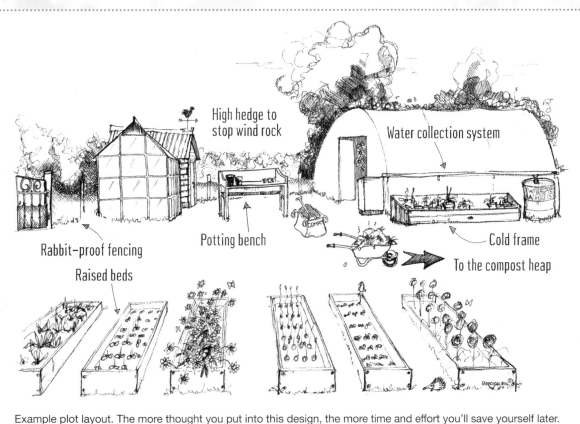

Example plot layout. The more thought you put into this design, the more time and effort you'll save yourself later.

Chapter two
Annuals

Annuals are your most abundant crops, giving instant satisfaction to the grower and great joy to the buyer. If you are an impatient character and fair-weather gardener looking for quick results, then annuals are for you. Which cosmos, sweet pea or chrysanthemum will you grow? Read on, be inspired, then hunker down with the seed catalogues and have some fun choosing.

Everlasting flowers are easy to grow and dry beautifully for winter colour.

Growing cut flowers, whether for pleasure or profit, is very much about creativity and taste. You can plant a field of annual asters to crop, but which asters will you choose? Will they be those listed in the wholesale seed catalogue as 'pink', or will you trawl through the Internet, spending time as well as money on finding a seed supplier for that one gorgeous annual aster you fell in love with when you saw it at a horticultural show?

Remember, you're competing with all the other growers, large and small, for market share. You will enjoy the whole experience a great deal more if you grow flowers you've chosen to match the garden you've planted: those which particularly tickle your floral taste buds – and they'll be a much easier sell for you if you love them. This may sound an obvious point to make, but when you look at the sheer quantity of different varieties of annuals, and then at the different colours within each variety. . . Which cosmos will you grow, and why? Which of all the sweet-pea varieties will you choose? The choices you make with your annuals will make your busiest season the success you want it to be. So in a small patch, avoid any kind of generic pink and go for unusual varieties.

You grow perennials for reliability, shrubs for filler and winter greenery, bulbs for early spring and autumn colour – but you grow annuals for delight and the sheer abundance of them. Flowering annuals provide the quickest turnaround in the gardening year, and with good management will give you buckets and buckets of crop from very little planting. Three cosmos plants, one sweet-pea stand, a little clary sage, some ammi, and a few tall snapdragon varieties *do* a cut-flower garden make – and from a planting of seed late March you'll be cropping that little mixture from June onwards.

If you plan to be a summertime-only flower farmer, then annuals are your crop: they're prolific; floriferous; come in every shape, colour, texture and height; and with the right management they'll perform for you all summer long.

How to grow annuals for cut flowers

Annuals are sold as the easiest of garden plants to grow from seed. That's because the goal of these plants is to make seed within a single growing season, and so they grow very quickly, flower quickly and try to produce seed quickly. This speedy turnaround makes them a satisfying project for the new-to-gardening enthusiast. Given the right conditions, they germinate in a matter of days, shoot up, and, if left to their own devices, can be in full flower in as little as six weeks, producing seed soon after and going over, job done, as fast as they flowered.

In a small patch, it's best to avoid any kind of generic pink and go for unusual varieties.

Your job is to make the most of your annuals by cutting their flowers and stopping the plants from setting seed. The more you cut, the more your annuals will flower for you. You should even consider (counter-intuitive though it always feels) pinching out the leading first buds of *all* your flowering annuals, to encourage side shoots and therefore more flowers (see pages 144-5, where this is described for sweet peas).

Not only do you need to create the right conditions for all this floriferousness, but you also need to manage your crops to suit your own requirements (after all, you are the boss). I'm a great fan of a little of everything: some direct-sown, some in trays, and staggered sowing for successional flowering and a longer season. You can never predict the weather in the UK, so hedging your bets is *the* route to success: whether you're growing acres of flowers for market or just a few for yourself and your friends.

When to sow

Pay attention to whether your annuals are hardy, half hardy or tender. Hardy annuals are plants grown from seed that can be sown in the late summer or early autumn, will shoot into a sturdy little plant, overwinter successfully unless the conditions are really extreme, and flower early the next year. Half-hardy annuals don't like frost and won't overwinter in a frosty area, but will survive a cold night or two in the spring. Tender annuals won't survive any frost, so shouldn't be planted out until all risk of frost is past. There is no point in direct-sowing a tender annual such as cosmos earlier than April, as it will be bitten off by the slightest threat of frost.

There are so many annuals to choose from that you may find that being held back by the weather helps you stagger your plantings into manageable chunks. A good way to manage your time is to plan to sow seed fortnightly from January until the end of May, starting with the incredibly hardy sweet peas, then pot marigolds, cornflowers, etc., and finishing with zinnias, cosmos and bells of Ireland, to flower in September and until the first frosts.

See pages 42 and 47-48 for more about sowing hardy, half-hardy and tender annuals.

Sow direct or in trays?

There are some annuals that really don't like having their roots damaged by being potted on or planted out, and so prefer to be sown direct. Larkspur, cornflowers, zinnias, gypsophila and ammi all do

Statice seedlings ready for potting on – or for planting out under a nice protection of fleece, if it's earlier than the end of April.

You grow perennials for reliability, shrubs for filler and winter greenery, bulbs for early spring and autumn colour – but you grow annuals for delight and the sheer abundance of them.

better this way – though you *can* start all of them in trays if you'd rather control the number of plants you'll have. It's a question of scale. If you want a 100m (330') strip of ammi, then you're probably best off sowing it direct. If you only want 15 plants in total, then it may be better to grow your seedlings in a tray. These are factors worth considering when planning your planting: thinking ahead will save you time and money later.

Neither half-hardy nor tender annuals will cope with any serious frost, so if planting early, you should plant them in trays under cover somewhere warm. But you can also save time by direct-sowing them later in the season: they'll get away quickly and make stronger plants if they don't have to go through the stress of being potted on from tray to pot and eventually to field. However, if you get a warm, wet spring, you may get a plague of slugs too, and they can take a whole crop of direct-sown cornflowers or cosmos in one night.

So again, hedge your bets – sow some seed directly and some in trays: if the direct-sown plants get away and survive the slugs, then you can give away or sell your tray-grown plants. If you lose a row of just-germinated gypsophila to a marauding army of evil molluscs, then you have spares to replace them with. Sounds labour-intensive? Well . .

Seedbeds and seed compost

Whether you're sowing outside into soil or into seed trays, for best germination the earth or compost you sow into should be worked into a fine, free-draining tilth which you can tamp down to prevent roots growing into air pockets, in which they would shrivel up and die.

This tilth is easier to create in the controlled environment of a seed tray than outdoors, but it's still worth doing if you're sowing direct. Repeated raking to break up lumps of soil, and using the back of the rake as a tamper, will make you a smooth, relatively air-pocket-free seedbed. Rubbing bagged compost between your hands into a bucket to break up the lumps and mixing with vermiculite to make a good seed-sowing medium for trays is much kinder to your back – but plants grown in trays will need potting on and planting out. Planting in seed trays may seem like less hard work when you're doing it, but you will have to handle those plants more, using more of your precious time, than if you direct-sow into a perfectly tilled seedbed. See page 56 for more about seed compost. Water seedlings in trays from underneath to stop seedlings flopping, drowning or damping off, and to prevent botrytis developing.

As long as your soil is weed-free, sowing direct is the least labour-intensive way of growing annuals outside. You can sow hardy annuals outside this way in September. If the autumn is generous, by the time the cold weather arrives you should have some good-quality, strong little plants. They may get battered and nipped by the winter cold and wind, and in the mean days of February look pretty unhappy, but underground their root systems will be well established and new shoots will get away quickly for you as the weather warms up in the spring. The nipping frost works in the same way as you pinching out leggy growth, and so autumn-sown hardy annuals are often really bushy plants, with lots of flowering shoots after being nipped through the winter months.

The right conditions

As always, first consider your earth. Feed it with a mulch of a nice, earth-replenishing layer of well-rotted horse manure or garden compost, add drainage (grit or sand) if your ground tends to get waterlogged, and add organic matter for moisture retention if your ground is too free-draining. Rake your bed to a fine tilth before sowing. If you're planning to direct-sow in the spring, wait until the ground is no longer chill to the touch: the old boys say you should sit bare-bottomed on the soil to test the warmth. . . I prefer to test with the back of my hand. It's amazing how you *do* feel the difference between chill and warmth in the soil that way. Hold your hand an inch above the soil and, if it's warm enough, you'll feel the heat radiating. Try it – I promise it works.

You should see me, from the beginning of March onwards: in the morning I step out into the garden with the dog and sniff the air for that green scent of spring, then look down to see what's poking through the dark mulch on the beds, and finally I'll squat down and hold my hand, palm down, flat above the earth, hoping for the winter chill to be gone. And eventually, sometimes as late as the first week of April, after all that faithful waiting, the day will arrive – and out will come the rakes, the heap of seed packets, the labels and the indelible pens, and the direct-sowing will begin.

How to sow

Direct-sow your seed sparingly in drills 30cm (1') apart. Don't waste your seed or think "I'll just use this packet up" and sow too thickly – you will have to thin your seedlings even if you sow very

A bank of cosmos flowering in October.

thinly. Once the seed has germinated, and that little row is a sharp green line along your drill, hold back before thinning out. The slugs will probably take some of your seedlings. Wait until you've got a row of well-established plantlets before thinning your seedlings to about 22cm (9") apart. Give your seedlings plenty of space. Think: a fully grown cosmos plant will be about 1.2m (4') high and the same in width so give it generous spacing or it'll fight with its neighbours. Cosmos plants especially don't like fighting; they just give up and you lose them.

Hardy annuals

Your first sowing of hardy annuals can be made in September, for an early crop the following year. If you're working a rotation in your cut-flower patch, then a good place to sow is where you have just lifted a spent biennial crop such as sweet William or foxgloves, for example. You could direct-sow half of your seed this way, and save the other half to sow either direct in the spring or in seed trays under cover in January or February. That way, if your luck fails and a flooding wet winter rots your autumn-sown plants, all is not lost.

Sowing direct

If you live in a gentle, warmish area where autumn tends to come late, you might want to consider holding back your autumn sowing until after the autumn equinox (23 September), because in a warm, sunny September, annual seeds sown much before that date may germinate and rush to flower before the end of the season, producing weedy little plants and going over before the cold weather comes – at which point you'll curse the time you wasted planting them and the money thrown away on the seed they grew from. Wait until after the autumn equinox and there will be less light than darkness during each 24-hour period, so the seed won't be tempted to flower before the season finishes. If you live further north, or on higher ground, or where an east wind picks up earlier in the autumn so that the days are cooler, then there's no need to worry about your plants flowering before next spring.

Hardy annuals may also be overwintered in trays in the greenhouse or polytunnel for a bit more protection, though they may still get frozen solid on cold nights. If this happens, they shouldn't be allowed to defrost too fast. Cover them with sacking and/or move them into the shade and they'll defrost slowly and will hopefully survive better than if suddenly melting in a sharp, glass-magnified winter sun.

Sow further crops of hardy annuals in the late winter and spring to take you through the summer. September-sown plants will not still be flowering the following September, so if you'd like a summer-long crop of calendula, cornflowers or sweet peas, or the unbelievably useful *Ammi majus*, then do sow a second crop in January, a third in March, and even, in a dry summer, a fourth crop at the end of May. Watch your weather: feel the sun and the moisture content of the earth – a hot spell in early July will push your early hardy-annual crop over quickly, so be prepared, spread your sowings, and all will be well.

As with all plants grown from seed, learn to recognize self-sown seedlings in your garden and you can move them to where you want them to grow on next year. While hoeing is great for quick weeding, if you suspect that there might be precious seedlings in the mix, then a hand-weed will save you time and money later on.

Hardy annuals I would not be without

I recommend this 'shortlist' of hardy annuals for a variety of reasons: they might be easy to grow, breathtaking to look at, have wonderful heady scent, or make gorgeous highlights in a cut-flower

Hardy annuals good for autumn direct-sowing

This is a group of what you might describe as the basic essentials in the hardy annuals world. If your September-sown seed survives flood, storms, frost, snow and any other unexpectedness that winter can fling at it, it will be flowering for you in May the next year.

- Ammi
- Cerinthe
- Cornflowers
- Larkspur
- Phacelia
- Sweet peas

A sea of ammi in front of a permanent border of verbena bonariensis.

bouquet. Think about all these different factors as you choose what to plant. How will your crop look, not in your patch but lined up in buckets ready for sale?

Ammi

If sown in succession, the lace-capped flowers **of** *Ammi majus* and *A. visnaga* will take the place of cow parsley for the cut-flower grower when the cow parsley is over, giving you lace edging for your flowers all summer long.

Sow your ammi in late autumn, around the end of October, and you may see no results until early spring, but then your crop will take you by surprise and get away amazingly quickly, providing an abundance of lacy froth as soon as the cow parsley is finished. To maintain an ammi crop through the summer you might sow fortnightly, or perhaps monthly from February (when you could start it in seed trays under cover) through to mid-June.

The heavier variety *Ammi visnaga* can be found at market, but the more delicate *A. majus* is more difficult to find, so this might be the one to grow if you're looking to supply local florists. There's an unusual pinky-purplish variety too.

Bupleurum

With teeny acid-green flowers on stalks of attractive fresh, round foliage, this is a very useful hardy annual. We use it as a filler from April to September, direct-sowing our first crop in September and successively sowing further small patches in February, March, April and May. It does the same job in cut flowers as alchemilla. If you sow it in small patches regularly through the spring, you can have it all summer long.

Cerinthe

The amazingly hardy *Cerinthe major* 'Purpurascens' can be bitten and bitten and bitten again by the frost, and it'll just come up bushier and more giving in the spring as a result. Its common name, honeywort, tells you how the bees feel about it.

The silvery-blue foliage – almost as squelchy-looking as a succulent – on beautiful arching stems, is as useful as the flowers. An autumn-sown cerinthe plant might be cropping as a filler as early as April, giving you welcome variety in greenery and texture after a long winter of pittosporum, eucalyptus and ivy.

Save the seed and make sure you plant it, as it can be expensive to buy. I lift my self-seeded plants and pop them into a patch, and there I have next year's crop. You'll quickly learn to recognize the seedlings because of the distinctive foliage. Cerinthe doesn't travel brilliantly out of water and I've never seen it in a traditional florist shop – it can sometimes flop after cutting, but revives nicely for me with a little searing.

Chrysanthemums

There are lots of different varieties of annual chrysanthemums, and I recommend you look at all of them and choose good colours for your mix. We grow the yellow 'Primrose Gem' to go with the spring colours, and the chocolate-and-mustard-centred, white-flowering 'Polar Star' for our late-season dahlia heavy bouquets. Very hardy for overwintering, it'll flower as early as May in a polytunnel, making a very useful early crop. When your customers ask for 'daisy' shapes, annual chrysanths are very useful.

Clary sage

We grow all the different colours of annual clary sage: dark blue, white and pinks. The spikes of coloured flower bracts make wonderful structures for bouquets, and cut into bunches of ten they look great in buckets for a market stall.

Grow lots of different-coloured hardy clary sage – useful throughout the season.

When you've had enough of tulips and other bulbs, the first cornflower is like a taste of a July cornfield – a promise of summer already flowering in late spring.

Annual clary is one of the hardest workers in the cut-flower garden: ours grows best for us in the tunnel. Cut it back after a first flowering and it will flower a second time for you on the same plant stock. We do plant successfully as well though, to ensure a really good-quality crop all summer long.

Cornflowers

Alongside poppies, ox-eye daisies and field marigolds, wild blue cornflowers are one of the definitive wild flowers of a cornfield mix. Cultivated varieties now come in blues, reds, whites, pinks and almost-black purples ('Black Ball'). They stand well out of water and dry well, so are hot favourites with the real-flower-confetti growers. They're often called 'bachelor's buttons', because tucked as a single flower through the buttonhole on a jacket, they'll keep smiling all day.

Sow cornflowers direct in September and you may well have a marvellously healthy crop as early as May the following year. You'll lose some plants to the weather over the winter, but others will bush up to fill the space. When you've had enough of tulips and other bulbs, the first cornflower is like a taste of a July cornfield – a promise of summer already flowering in spring.

We don't grow acres of them. But I do sow cornflowers successively, and have a little patch in

Cornflowers come in many different colours: this one is 'Black Ball'.

flower all summer long. They're not the most valuable cut flower to grow, because their heads are small and their stems can be weedy, But they are one of the easiest and, certainly if you're thinking of gate sales or farmers' markets, bunches of cornflowers will always do well for you.

Larkspur

These annual delphiniums, with their tall, papery flowered spikes in whites, blues, pinks and reds, are hot favourites with florists, and will crop reliably over a good two-month period from a single sowing. They are much grown by specialists in making real-flower confetti, for fresh and dried use. If well conditioned, the flowers don't wilt quickly out of water, so are useful in hair garlands and buttonholes.

Larkspur do particularly well planted out in September to survive a sharp winter, as a few weeks of freezing temperatures will help with vernalization of the seed, so ensuring a good germination rate. (Vernalization is the encouragement of germination, in the case of seed, or flowering, in the case of perennial plants, as a result of exposure to sufficient cold.) Larkspur seed should always be kept in the fridge once you have it, and if you're sowing larkspur in spring, so it won't get any cold treatment outdoors, then do put it in the freezer for a couple of weeks before sowing to it help it germinate.

Nigella

Nigella ('love-in-a-mist') overwinters well, and a first crop will flower as early as May. Its gorgeous spiky flowers are a magnet for bees, and the seedheads make great additions to floristry later in the season. Plant a second crop in February for successional flowering. The tallest variety is the classic 'Miss Jekyll', but there are whites with moody dark centres, and pretty pink varieties too. Keep different varieties well apart, because they will mix, and the self-seeded plants next season can't be guaranteed to be the colour you want.

Orlaya

A magnificent lace-cap, less often found than ammi and with a heavier lace design in the flower head, orlaya overwinters well and will flower in May. The seed is expensive, but saved seed will come true easily. We find this grows better for us direct-sown in beds than brought on as seedlings in trays, so we successionally plant small patches of orlaya in September, January, March and even again in May, to ensure a summer-long supply.

Phacelia

Often planted as a green manure, the flowers of phacelia are beautiful in May and June and are usually humming with bees as soon as they come out. An especially intense purplish-blue, they are extremely useful as a cut flower. The stems are strong and the flowers hold well in water and are large and fascinating to look at. It will self-seed all about, but seedlings can easily be managed into a block if you keep an eye out for them.

Pot marigolds

If you plant pot marigolds once in your garden, they will inevitably self-seed and you'll have them forever. We grow several different kinds – my favourite is the silvery petalled 'Sherbet Fizz'. I love the slightly herby scent of pot marigolds when you cut them, and, while they might look as though they'll wilt in a heartbeat, cut directly into water they condition well and last for up to a week. Be careful when cutting them, as their stems can be brittle and easily snap. You'll have to get used to stripping their foliage gently. They're very hardy and are one of the first hardy annuals to flower in the spring if overwintered.

Quaking grass

This is a sparkling little light in a posy or bouquet. We don't grow masses of it, because its delicacy makes it quite labour-intensive to cut for a not-amazing return. But it's very pretty for brides' bouquets and dries nicely for Christmas wreathing.

A combination of phacelia and pot marigold is as pleasing in the garden as it is in the vase.

It will self-seed everywhere if not cut ruthlessly, so grow it if you really love it, but not if you're going to forget it.

Scabious

Annual scabious is quick to germinate and easy to grow. We make sure to grow the fascinating stellata variety 'Drumstick' every year for the gorgeous structural seedheads, as well as some of the pretty cottage-garden-style varieties such as 'Snow Maiden', which is brilliant for weddings and, because it's slow to wilt, for buttonholes, corsages and bridal crowns. Keep cutting it and annual scabious will flower and re-flower all summer, so it's good value for the space it uses.

Snapdragons

It's worth thinking about growing several successional crops of snapdragons. We plant our first crop as seed in September, and plant them out in the polytunnel as soon as there's room in late autumn. We sow more in January and March, and direct-sow another crop in early May. They're hardy, lush-looking, and make lovely tall spikes for floristry. Your local florist will be able to get snapdragons from their wholesaler easily all year round, but if you're growing for your own market stall, you should choose some lovely colours to go with your scheme and grow lots of them. They will re-flower from side shoots. While these side-shoot-flowering stems might be a little delicate for selling one by one, if you're doing your own floristry they work well in small posies.

Sweet peas

Synonymous with spring, with scented cut flowers, with what we grow best in the UK, sweet peas (in case you haven't noticed yet!) are my out-and-out favourite cut flower (see Chapter 9 for much more). For value versus time spent growing and

nurturing, etc., they can't be beat. You can charge as much as 65p a stem for a sweet pea (50cm/1'8" long), and if you don't treat them with silver nitrate, then the scent will have your customers swooning.

Although they are amazingly hardy, you might think twice about direct-sowing sweet peas in the autumn if you want to grow long stems to cut for sale. Sharp frosts might pinch out your plants so often that you'll end up with five or six flowering shoots per plant. While this will give you bushy, healthy plants with great root systems, you probably won't get the stem length you'll need if you're planning to sell top-quality sweet peas to florists or direct to customers.

Sow sweet peas in deep seed trays, or even in toilet rolls (though see page 143), which you can keep in a cold greenhouse or protect from the worst of the frosts until you plant them out in late March or April. Remember that sweet-pea seed is also the mouse's favourite supper: sweet peas sown direct in September may be mouse-munched before they've got much past germination. The seed is (relatively) expensive, so I don't waste money direct-sowing sweet peas at any time of year.

Half-hardy and tender annuals

You can sow half-hardy annuals in trays under cover from early March. If you want to sow them direct, you might want to wait until April. You will learn which ones do well for you and which to avoid sowing too early. Because they are annual seeds, given the right conditions, they'll grow away quickly.

Don't start sowing your tender annuals, even under cover, until the beginning of April at the earliest. Because you can't plant them out until all threat of frost is gone, you don't want them getting too big under cover and needing to be potted on again and again. Tender annuals can be planted as late as mid-June for a late-autumn crop: you may not be rewarded with the plant height you'll get from earlier sowings, but you'll get a good show at the end of the summer. If you're sending flowers to market, you'll need longer stems than if you're growing for your own use, for farmers' markets or for gate sales. But I still think a late sowing of half-hardy and tender annuals is definitely worth the cost in shorter stems. You might have space where your tulips have just come out, for example, in which this late crop could be direct-sown.

If sowing before the danger of frost is past, both half-hardy and tender annuals should be sown in seed trays under cover somewhere warm. It's at this time of year that you begin to see eager gardeners who don't have enough greenhouse space filling all the available window sills in their houses with tender seeds in trays. The trays sprout quickly and the annuals can soon look weak, leggy and lopsided as they reach for the light. If you can't wait until it's warmer outside, then by all means sow in trays and keep the seedlings indoors – but make sure they're somewhere light enough for them to grow straight and true. A simple heated propagating bed, such as a warm cable through a sand box (see Chapter 1, page 30), in a greenhouse is a better place to start your seedlings off than on a window sill, as the light will hit them more evenly. The warm cable will prevent the air around them from freezing, even if other parts of the greenhouse get very cold.

Don't plant out your half-hardy or tender annuals until all risk of frost is past. If you've sown some hardy annuals in September, you'll have plenty of these flowering for you throughout the spring and early summer, and you won't need your half-hardies and tenders to flower until late summer – so don't rush them into the ground when a frost might nip them and undo all your hard work. You could hold the seedlings back and not plant them out until the second half of May. Here at Common Farm Flowers we never plant out any of our tender plants until 1 June. If you're organized, you won't need to either, because you'll have a carefully planned cropping schedule (see Appendix 1 for an example), which will be filling your flower buckets in spring and early summer with hardy annuals, biennials and a lovely crop of perennials and useful early shrubs. (Yes, a spreadsheet is useful if you want to grow a wide variety of cut flowers!)

This malope, *Malope trifida*, has interesting seedheads as well as beautiful, if easily bruised, flowers.

With careful planning, you can stagger your planting in a more manageable way. You're also clearing your beds of old crops at the right time for a new crop to go in, and you don't have a great many plants waiting about in pots queueing for bed space and increasing your water bill.

Half-hardy annuals I recommend

Your half-hardy annual crop will start flowering from July, so pick those that go with your dahlias and late-summer perennials. There is such a wealth of plants to choose from for cut flowers that it's always good to remember you have a colour scheme of your own, in order to help you edit your choice.

Asters

Not enormously hardy, annual asters will grow stronger and be happier if not disturbed, so direct-sow them in April and thin out the seedlings, rather than growing in trays and planting them out in their growing positions. There are some really marvellous colours and shapes in the annual aster list, so look carefully. They make good, showy flowers which should sell well per stem, as well as make real impact in your bouquets. My favourite is 'Florette Champagne', not always easy seed to find.

Bells of Ireland

Again, these seem to grow much better for us when sown direct in their flowering position. They are absolutely worth growing: their tall spikes of green flowers make a wonderful accompaniment to late-season dahlia bouquets, will stand tall in big wedding arrangements, and are great in floristry when the customer's looking for structure and architecture in their flowers. They

can be difficult to condition, so be sure to cut them early in the morning and give them plenty of conditioning time to recover from the trauma of being cut. Also, don't be surprised when cutting them if you feel a sharp prickle: for such lush-looking flowers they have incongruous teeny little thorns beneath the flowers, which can hurt if you're not expecting them.

Cleome

This is on the recommended list because everybody else loves it – but not I! For me, it has a foul stink, and the sharp thorns under the leaves are vicious compared with the gentle prick of a bells of Ireland thorn. The flower heads are interesting, but I find them too loose for use in bouquets, though they are great for big arrangements in

Due to its spines and its horrible stink, cleome is my least favourite flowering annual! But it may be your most beloved.

pedestals or for church window sills. There are good colours – pinks and whites – but cleome is included in this list more to draw your attention to it than because I'm saying you ought to grow it. As I say often in this book, your plot is for the flowers you particularly like.

Craspedia

This gives you a good drumstick of interesting rich mustard-yellow; it's great out of water, and I love the colour. Craspedia does a similar job in floristry as thrift, only is quite a useful yellow. Some people are very anti-yellow in a scheme, but I feel it lifts what is often a pink, blue or white mix and so am an enthusiast for plenty of yellow in a cut flower scheme.

Didiscus

I love these delicate, lacy flowers. Not unlike scabious, perhaps a little looser in design, didiscus are a challenge for me and I've yet to grow a successful large crop. This year we're direct-sowing

them late (April), and so far they're germinating nicely in the field. But it's been a warm, early spring, and we may yet get a hard frost, which might nip them all off if I don't get out there with the horticultural fleece in time. If they grew very easily for me, I might prefer them to some of the annual scabious varieties (though I'll always grow the stellata drumstick scabious).

Mallows

These outrageous flower trumpets are extremely delicate and prone to bruising at the slightest touch. They are, however, certainly showy enough to grow for your own use. I love the seedheads as well, and have been known to pull off bruised petals to keep the seedhead inside for floristry.

Nasturtiums

Nasturtiums are surprisingly good cut flowers, as well as good salad toppers! They are very popular for sale as edible flowers, but so long as you get the stems straight into water, they also work as

Nasturtiums make good cut flowers – though don't ask them to stand for any time out of water.

short-lived cut flowers: long, trailing ends of nasturtium make a great addition to arrangements. Recommended if you're doing your own floristry, but perhaps not one to sell in bunches at market.

They're also great for bringing in pollinators and as a sacrificial plant next to cabbages.

Rudbeckias

Annual rudbeckias are generally described as half-hardy (though we've overwintered them here in the south-west of England), and so I'm listing them as such here. They last up to two weeks as cut flowers, and have sturdy stems and attractive, daisy-shaped flowers. The dark colours, such as 'Cherry Brandy', are useful at the end of the season, as are the more obvious cheery yellows. If sunflowers are too big for you to want to handle, then rudbeckias do a good job as an equivalent choice for colour.

Statice

Statice is worth growing because it doesn't wilt, has lovely colour variations, and dries beautifully for Christmas decoration work. Look for unusual colours: as well as the better-known bright purples and blues, there are good whites and a very pretty variety called 'Sunset' with a mix of sherbet colours.

Verbena bonariensis

We treat this as a perennial too at Common Farm (see Chapter 4, page 81), and it isn't killed by the winter so long as we leave the previous season's dry stems to protect new growth from frost. Without protection the plants wouldn't overwinter. Here in Somerset our winters are really very mild, so I'm listing verbena bonariensis as a half-hardy annual, because elsewhere you might find it easier to grow from seed at first – though in subsequent

years you will very likely find seedlings popping up all over the place if you don't cut every single flowering stem. To grow from seed you should treat it as half-hardy and sow under cover in March. The seeds are minute and the seedlings minuscule, and will benefit from a quick pricking out so they have room to develop in a pot before being planted out as a good-sized plantlet in May.

For good vase life, cut the flowers before they're fully open.

Good tender annuals to grow

Again, take time to curate your choice. Bear in mind the colours of your dahlias, and of the turning autumn leaves with which your late-summer annuals will be arranged, and choose accordingly.

Amaranthus

We direct-sow our amaranthus ('love-lies-bleeding') at the beginning of May. The long trails of developing seedheads give an amusing added dimension to cut-flower combinations. Sold by the stem or as part of bouquets, it's certainly attention-grabbing. We grow the green, pink and red varieties.

Cosmos

This is a classic to grow for cut-flower production: it's prolific, it comes in many colours, and it provides a lovely, light, fresh daisy shape through the second half of the summer and until the first frosts. Cosmos won't travel well out of water or laid flat in boxes, so it's good to grow for the local market. Choose colours from white through pinks to oranges and reds. Choose different petal shapes. Choose semi-doubles. There are so many varieties, so take your pick.

Everlasting flowers

Very useful in late summer bouquets, everlasting flowers sell well at market because they are just that, everlasting. And they're easy to dry for Christmas decoration. The first time I grew them I was surprised at how much I liked them as a fresh cut flower, and in the end used them much more that way than I did dried (which was what I'd grown them for). They come in lovely, really bright, jewel colours: great with dahlias.

Nicotianas

The seed of nicotiana is so tiny that we don't direct-sow it, because you inevitably sow in too-close-together clumps. In trays it germinates quite quickly, and benefits from a really speedy pricking out. You might not think nicotiana would make a brilliant cut flower, as it looks a little as if it would wilt, but in fact it cuts very well, and stands for up to a week in water. Look for interesting colours (my favourite is the lime green). Remember to check that you're ordering seed for taller varieties, not for dwarf bedding.

A bunch of everlasting flowers cut for drying.

We don't grow pollen-free sunflowers because we need to feed our bees as well as grow cut flowers for money.

Sunflowers

We grow sunflowers in seed trays, because, though they sulk when being planted out, their seed is too easily lost to a marauding mouse in the field. Besides, their eventual size means it's more economical to plant seedlings out at a good spacing than to sow direct and then thin. Pinch out the budding tips of your sunflowers to encourage side branches, producing many more smaller and more useful flowers. And look for the chocolate- and lemon-coloured varieties: you don't have to just grow plain yellow. Depending on your love for bees, you might also look for pollen-free varieties to grow for your market. Bees mow the pollen off sunflowers outside, so you may not realize how much they produce: pollen which will develop and dust the area around a sunflower bouquet indoors, potentially irritating your customer. We do not grow pollen-free varieties of sunflower, because we believe in paying a tithe to nature (see Chapter 11).

Zinnias

Direct-sow if you can, because zinnias will sulk if you disturb their roots. Their lovely sunny faces make a good, reliable, sellable crop for late summer. Order seed for the Benary's Giant series: again, you don't want to find you've grown a lush crop of dwarf bedding plants. Their colours are all delicious, so choose to match your other cut flowers and enjoy them.

Where to get your seed from

I save seed, I buy seed, I am a seed-aholic. My precious brown paper bags of saved seeds are kept in a box in the cool of my office (always cold and good for seed-holding), and next to them are the packets of bought seed, which comes and goes more often, ordered in little tranches throughout the year. Trawling through seed catalogues and choosing the promise of the next season has to be one of the treats of growing cut flowers, whether for pleasure or profit. I even add this happy necessity to my diary. Seed buying is a welcome opportunity to discover new things, to put into practice what one has learned, or to develop an idea that may have been budding throughout the current year. Seed is cheap and

---

WHAT WE'VE LEARNED

Seed catalogues can be naughty and list seed as hardy in order to make sales. Sunflowers are not hardy – they will not overwinter – and yet they're often listed as such by seed suppliers.

exciting – the perfect combination for a gardening shopper!

Saving seed

Annual flower plants will produce seed in abundance if you let them, so there is every reason to make the most of this and save on your seed bill for next year. But you might consider growing the plants you want to harvest seed from in a separate patch: this way you can leave them to go over earlier and can allow for the inexact nature of the weather generally, giving the seed longer to ripen. If you crop and crop and crop a plant until just before the first frosts, you can't expect it to produce fat, healthy, ripe seed. If you let one or two plants go over earlier in the season, however, you can harvest good, sun-fattened seed in September, before it's washed into the ground by the wet autumn weather.

Saving seed is as simple as sowing it. Bundles of brown paper bags are inexpensive when bought online, and a pack of 500 or so (the quantity they're sold in) will last you a good long while – and let's face it, is there a household anywhere that doesn't need a bundle of brown paper bags for a great many more reasons than collecting seed? Keep an eye on your seedheads and crop them before they explode all over the ground – nigella will do this more quickly than you might think; ditto poppies.

Make sure you label your bags of seeds carefully when you harvest them. You may think that you'll never mistake nigella seed for anything else, given that it's in those amazing architectural seedheads which look like mad punk haircuts on green-and-purple-striped faces. But you may be growing three or four kinds of nigella, and will need to know which is which. And even if you're only

Grow nigella for flowers and gorgeous seed pods – it will self-seed ruthlessly. . .

collecting one kind, when it comes to sowing time you may find that those distinctive seedheads have crumbled to dust and, unless you know your seed well, you won't be able to identify it.

Harvested seed should be kept somewhere cool, dry and out of direct sunlight until you're ready to use it. As with any bought seed, you should use it quickly: the sooner it's sown, the better the germination rate.

Buying seed

There are almost as many seed-supply companies as there are kinds of seed. You should try them all. Buying wholesale will give you a lot of seed cheaply – but think twice before buying by weight. Retail-sized packets of ammi seed, for example, give you 500-plus seeds, which weigh practically nothing. You should get a pretty good germination rate from bought seed (the seed merchant shouldn't be selling you anything that, if planted properly, won't give you about 90-per-cent success) – so why would you need a great deal *more* ammi seed than you get in a retail-sized packet? Buying seed by the gram will give you way more than you're likely to be able to use up while it's fresh, especially if you're growing on an allotment-sized or smaller plot. So, your seed-buying habits will very likely be dictated by your available space, as well as available time and finance.

With seed, the one thing you should really be concerned about is freshness. Stale seed won't have a high germination rate and will have cost you time and money for little return. Buy direct from suppliers (they all have very good websites you can buy direct from) and you'll get good-quality, fresh seed.

Don't leave seed-buying till the last minute: you want to get *your* choices, not everybody else's

leftovers. There's nothing more annoying than seeing 'sold out' under the best of the list you've made for next season. On the other hand, don't buy too early either. Buy seed when you need it: order hardy annual seed in August for your September sowing, then again in December to start planting in January. Order half-hardy and tender annual seed in February and March, and biennial and perennial seed in April or May for a June planting.

What you'll need for growing from seed

It's worth checking your seed-planting kit before starting. A tidy, well-supplied potting bench will allow you to plant seed quickly and efficiently: you don't want to be endlessly popping back to the garden centre to buy trays or labels or pens. . .

A note on wholesalers: unless you're buying a great many trays/pens/labels, etc., you may find that your local garden centre is no more expensive than buying wholesale, and when you're starting out, you won't need to spend so much on buying so many. Always double-check prices – just because a firm 'supplies to the trade' doesn't necessarily mean they do it at competitive prices.

Seed trays

Second-hand seed trays are good, but make sure they've had a quick wash, with disinfectant, to get rid of slug eggs and mould.

Use deep seed trays for sweet-pea seeds: leguminous plants have long roots and they'll do better if the roots have space to rootle down into.

You'll also need large shallow trays (without drainage holes) into which your newly planted seed

With seed, the one thing you should really be concerned about is freshness.

trays can be put to absorb water from underneath. Never water seed trays from above: let the compost soak up what it needs from underneath, once or twice a week. Watering from above encourages damping off and may wash away your carefully planted seed so that it all sprouts in a difficult-to-prick-out mass in one corner of an otherwise empty seed tray. Check the surface of the compost in your seed trays, and water them if it feels dry.

Labels and pens

When labelling your sowings, remember to mark the name of the seed supplier as well as the variety and the date of the planting. You'll quickly see whose seed you like best, not only from the varieties but also from the quality of the germination rate.

Large, plastic labels are best marked with an indelible pen. Labels are often sold with pens that are not indelible, and it is irritating when you realize this too late, when your watering regime has washed away the information on your labels!) We buy indelible pens – the sort you can use to write on CDs – from our local stationers.

Compost and vermiculite

If sowing seed in trays, do use an organic, peat-free compost: it is environmentally unsustainable to use peat-based compost, so do, please, try using something else. Bought peat-free compost can be of varying quality, so it may be worth modifying it. We use sieved municipal green-waste compost mixed with about a quarter as much again of sharp sand for drainage, plus a few handfuls of vermiculite per barrowful to help with water retention (we like to cover all possibilities), and it does very well for us.

We use this same mix of compost when pricking out, but when potting on we generally use just municipal green-waste compost, not even necessarily sieved (we just pick out any big woody bits), perhaps with a handful of sand if it's very wet. When making your own compost you'll quickly get a feel for it – as you run your hands through it, you'll know whether a little sand is necessary, or whether your plants, which might be going to sit in this compost for only two or three weeks before being planted out, will be fine without.

You could mix in a little coir compost too, but we find coir very hard on the hands and not brilliant for retaining water. It is light, however, and if you're carrying a lot of trays about then you can considerably lighten the load by mixing coir compost into your seed mix.

For seed sowing, shop-bought compost is better than home-made. It will have been sterilized (the same should be true of municipal green-waste compost) and so should be free of weed seeds as well as mould and pathogens. In the confined, sometimes airless conditions of a winter greenhouse or polytunnel, mould and pathogens can develop into botrytis and rot off your seedlings very quickly. Home-made compost may harbour slug eggs and mould, neither of which you want near a freshly shooting seedling.

For slug wars

Vaseline and cheap table salt will stand you in good stead: a smear of Vaseline around a seed tray deters slugs, and a smear of salted Vaseline around a seed try will be twice as effective.

Coffee grounds, eggshells, grit, sand. . . If direct-sowing into clay soil like ours, you may (or rather will) find slugs a problem. We scatter baked egg-shells and spent coffee grounds on to the surface of our beds, and we're conscious that the wooden edges of our raised beds can be a day-haven for this number-one pest. I will admit that I go on murderous slug patrols at night. You can put slug traps at the ends of your beds – tin cans with a tempting, drowning pond of beer at the bottom; or leave upturned half-orange or half-grapefruit skins, under which you hope slugs will hide during the day so that you can catch and kill them. The successful flower grower is not squeamish about murdering molluscs.

Watch out for slugs hiding out in the wooden edging of a direct-sown annual bed like this.

Top five tips for growing annuals for cutting

❋ **Hedge your bets:** sowing half your seed direct and half in seed trays will mean that whatever the weather, you'll have a crop of annuals to sell.

❋ **Think financially:** just because an annual is easy to grow doesn't mean it's the easiest cut flower to sell. Before you sow, or order seed, think about whether the resulting crop is going to find a financially rewarding market.

❋ **Curate your choice:** there's such a huge variety of flowering annuals to choose from. Remember that your flowers are going to be sold together, so think before you plant too

blinding a combination (unless blinding combinations are what you're best at selling).

❋ **Remember that you're growing to cut your flowers, not just to look at them:** too many annuals to cut and you won't keep up with them – your plants will quickly go to seed and your crop will be over before you've really got started. Grow fewer plants in your first few years and be sure to keep up with the cutting, to ensure a good crop all season long.

❋ **Do save seed:** annual seed will be one of your biggest expenses. Saved seed will save you money.

Chapter three

Biennials

Remember to sow biennial seed in early summer, carefully nurture the little plants through the hot weather, and you'll have stolen a march on the next season – meaning you're not so dependent on your September-sown hardy annuals flowering early the next year. Sweet Williams, sweet rocket, foxgloves. . . These are staples of the cut-flower patch – key players in your bouquets in late spring.

With all the spring planting, the direct sowing, the just-flowering hardy annuals, the bulbs finishing, the roses budding up, the foliage in all its fresh green glory – not to mention the hard work cropping flowers from your garden, arranging and selling them, earning enough money to take you through the barren days of winter – it's easy to forget one of the major jobs of the gardening year. Biennials are a necessity in the flower-farming year: without them your garden next year will feel half dressed. There'll be a hole where the sweet rocket should be, there'll be no wallflowers, no foxgloves, no honesty, no California poppies. . . The list is long.

By early June you'll feel as though you've been planting out for months (you *will* have been planting out for months) . . . and with the dahlias out of their back-breaking pots and finally in their flowering positions, and the last of the late-flowering annuals direct-sown in the remaining few square inches of bed space you have, the last thing you'll be in the mood for is sowing more seeds. What's more, unless you have lots of unused bed space (unlikely, even in the largest of gardens), you'll have to sow these seeds in trays, look after them through the heat of the summer months (groan), and prick out the seedlings to grow on, ready to be put out in September.

But the satisfaction of having large, healthy plants to fill the gaps as you clear your annuals in late summer cannot be exaggerated. A flower farm with bare earth is a flower farm not working hard enough for you. A bed emptied of a spent sweet-pea crop one day and re-filled with next year's sweet William the next will give you a well-deserved glow, especially as those sweet William plants will work doubly hard by acting as a weed-suppressing carpet for you – because weeds like to germinate in September as much as they do at any other temperate time of year.

Sweet William is worth quite a lot as a cut flower because the big flower heads make good statement flowers in a bouquet. This is my favourite, 'Auricula-Eyed Mixed'.

Growing cut flowers for money is a satisfying, beautiful, creative and often exciting job – but it's also repetitive, and requires concentration and the will to keep on sowing and planting when all you want is to sit back and admire the fruit of your labours. No time! Keep going. Work you do on biennials in June will save you from panicking

Towards the end of summer, beds are cleared and mulched, ready for biennials to be planted out.

September: time to start planting next spring's biennials in their flowering position.

about being low on stock next spring. Making a success of growing cut flowers for profit is so much about planning seasons in advance.

How to grow biennials for cut flowers

Biennials are plants that flower in their second season. In their first year they grow vegetatively (i.e. leaves only), then in their second year they flower. Sow them in early June for producing well-established seedlings to plant out in September, for a flowering season the following May/June.

Remember to order biennial seed in spring (along with your perennial seed) and put the packets somewhere where you'll see them in early June. Since it is in that first fortnight in June, before the longest day of the year, that your biennials need to be sown in seed trays. They will sprout

and grow the fastest of all the seeds you sow throughout the year.

You need to sow them in seed trays because your beds will be full of annuals at this stage. Make your seed compost in the same way that you would for annuals (see Chapter 1), but perhaps include less sand and a little more vermiculite (you don't want your compost to dry out). You won't necessarily need a greenhouse to encourage germination, and certainly not a heated propagating bed. Your biennials will need pricking out quickly: foxglove seed especially is so small it's like dust, almost impossible to sow sparingly, so needs quite a speedy prick-out after germination.

You must keep an eye on those seedlings. Keep them from drying out, as they'll also fail the fastest of any seeds you sow throughout the year. They grow so quickly and hate being tray-bound or pot-bound: if that happens, they'll struggle to recover. Leave them for a fortnight and they'll sulk, drily. You might find you have to pot them on two or three times before there's space in your beds to plant them out in September. So don't forget to water, prick out and pot on your biennials. They'll reward you by flowering their hearts out the following May and June.

Which biennials to grow?

Depending on the conditions in your garden, many biennials will work as short-lived perennials – here in the south-west of England they certainly do. Some can be sown as late-flowering annuals: sow in spring for a crop in late summer. However, I'm a great fan of dividing the year's jobs into manageable chunks, as well as constantly ringing

WHAT WE'VE LEARNED

For the sake of my sanity, a manageable job list, and the limited space I can plant in, I can live without sweet William in September, when I'll have plenty of dahlias to keep my bouquets looking fresh. So now I don't sow a second crop of it to flower in late summer as an annual.

the changes in the crops I'm able to offer my customers, and so I sow biennials as a single job in early June.

The striking foliage of sweet William 'Sooty' can be cut as a filler before the flowers come – if you've planted enough of it, that is.

Recommended biennials

There are a great many varieties of the plants in the list that follows, and my preferred biennials list is by no means exhaustive, so do look at all the options available. As I've said elsewhere, your cut-flower patch may contain many of the same *sorts* of flowers as mine, but the chances are you'll choose different colours and different textures, and thereby create an entirely different look from mine.

California poppies

We grow these as biennials *and* as hardy annuals, sowing one crop in early June and a second crop in September. They're not so fast-growing that a June-sown crop will flower weakly before the end of the season (wasting the seed), but sown in the polytunnel it will give us a first flush as early as April the next year, then the September-sown seed (outside) will give us another crop in June. Sow in the spring and they'll behave like annuals, flowering later in the summer for you.

Despite their apparent delicacy, and the fact that they're known as poppies (though they're no relation of the *Papaver* sort), California poppies will stand happily in a vase for up to a week, dropping their petals rather than wilting, and making an interesting green arrow of a seedhead in the process. They come in so many interesting colours, from the classic blinding yellow-orange of 'Aurantiaca' to the rich pink of 'Rose Chiffon' and the subtle, paler, pink-and-cream mix of 'Champagne and Roses', which is great for weddings.

For cutting and selling by the bunch to florists you may have to experiment with keeping them under control – they can be wayward and too curly for a high-street florist's taste. But if you're making posies or bouquets yourself to sell, then I recommend them highly. Their lovely foliage, shaped like green coral, is also useful in a bouquet – an interesting change from ordinary leaf shapes.

Foxgloves

There's a great deal of work being done on developing foxglove strains, and many of those now available can be classed as perennials. If you're growing herbaceous borders in which there's room for experimentation, then by all means pop foxgloves in and see whether they'll come back year-on-year. If you're growing flowers for cutting, though, you might want to focus on growing foxgloves as a biennial crop. Seed sown in early June makes good, strong plants to plant out in September, which will give you a very useful crop in May and June the following year. If space permits, you can always leave your plants in the ground (as we do) to see if they'll re-flower the following year.

Foxgloves are really useful as wedding flowers. They will shoot from side shoots and give you a more delicate second cut. However, I think they're quite space-greedy for the crop you get from them, and are a bit fussy. In the past we've only grown a few, but I'm finding them increasingly interesting structurally and in all their lovely different colours, so I've already identified a huge bed we'll fill with them next year. They do like a

Don't forget to water, prick out and pot on your biennials. They'll reward you by flowering their hearts out the following May and June.

Foxglove 'Pam's Choice' is beautiful, tall and repeat-flowering – a great spring favourite at Common Farm.

leaf-mouldy bed and a protected, if not shady, environment, and they can also tend to rust.

Honesty

Honesty provides welcome early-season purple and white scented flower spikes. It doesn't last for ever in a vase, but the petals drop prettily and the stems don't wilt if well conditioned. Then those plants that you don't cut in May for bouquets at that time of year will make those gorgeous silver-coin-shaped seedheads so invaluable in Christmas wreaths and garlanding.

Sweet rocket

This has to be my favourite biennial, with its tall, strong explosion of scent and delicate heads of white or purple flowers, which will, in a less managed garden than mine, set seed about: the plant performing like a reliable perennial. Its scent is like honey, and its first flowers arrive in

time to be mixed in with the last of the cowslips, with cow parsley, forget-me-nots and late apple blossom, with black tulips. . . Are you sold on it yet? The stems' height and strength make them extremely useful for large wedding-flower arrangements as well as for posies.

When I give up flower farming, sweet rocket will still have pride of place in my borders. It will stand for over a week in even dirty water, and you can cut the heads when there's barely colour in them and still the flowers come slowly out.

If you're thinking of what jobs different flowers do in bouquets through the season, then sweet rocket takes over where lilac leaves off, and fills the gap until the phlox and panicle hydrangeas start flowering. Your first cut will be of the big, heavy heads, then the plants will keep sending up flowering side shoots of increasing delicacy, from which you will be cutting throughout the early summer.

Established plants will last a couple of years. We sow new ones every year, but keep the old plants too, moving them to a new place as we rotate our cut-flower crops about the garden. They're not hard work to move: at the end of the season we cut back the remaining flowering stalks and the plants lift quite easily, as they don't have especially deep roots. We also keep an eye out for seedlings when we're weeding. Sweet rocket is almost my favourite flower (after sweet peas . . .), so I keep them.

Sweet Williams

Cousin of those old-fashioned garden pinks, the sweet William is a true stalwart of the cut-flower garden. It is cheap to buy as seed and easy to grow. The flowers are great for selling to market, and their stem prices are always good because the flower heads are large and so make a big noise in a bouquet. There are so many varieties that you really can choose to fit your colour scheme. Look for green varieties such as 'Green Trick', and the very dark red 'Sooty', for example . . . although my true favourite is the classic 'Auricula-Eyed Mix' (pictured on page 60): a lovely mix of bicolour flowers, of which a 3m x 1m bed will keep you in sweet William for three months from May.

Thrift

Native to the wild sandy edges of land next the sea, thrift likes dry conditions but doesn't mind any quantity of wind. While not a huge-headed flower worth a great deal to sell by the stem, it is nonetheless extremely useful for the wedding florist: with lovely strong stems and small, neat flower heads, it doesn't wilt, which makes it great for hair flowers or buttonholes. There are lots of varieties (though all variations on pink), so look out for one that suits your colour scheme.

Wallflowers

Wallflowers don't last for a very long time as a cut flower, but if you're doing wedding flowers, their peppery honey scent will warm a whole room. Check that you're ordering seed for tall varieties, not dwarf bedding, and don't forget to stick your nose into the first flowers for the scent of spring. The colours can provide a lovely rich jewel extra in a spring mix that might otherwise be more muted – very useful.

Five top tips for growing biennials for cutting

❀ Remember to sow them! In the madness of a flowery early summer, the last thing on your mind will be sowing seed. Sow your biennial seed in trays in early June and you'll have big, strong seedlings to plant out when clearing annual beds in September.

❀ Remember to water them! Keep an eye on your biennial trays. They'll germinate quickly and grow fast through warm summer days. Don't let your seed trays dry out, and prick out and pot on your seedlings when they're ready, so that you have the best-quality plants to put out in late summer.

❀ Remember to make space for them. September may feel like an odd month for planting out, but if you've brought on several hundred biennial plants, then do schedule the time and space for getting them into beds for the winter. The results the following May will be well worth the effort.

❀ Don't order your biennial seed in the winter or previous autumn when ordering annual seed. The point of growing annuals and biennials for cut flowers is that you can cater to changing tastes easily – and your tastes will change as often as your clients'. Watch out for new and different varieties, and order seed when you're going to use it – then it'll match your taste at that time.

❀ Remember that some biennials will behave like short-lived perennials, lasting longer than one season: you can move them about to fit next year's planting plan, or keep them where they are if they're not in the way. The stems will be less vigorous in years two and three, but you'll still get a good crop.

Chapter four
Perennials

It is your choice of perennials that will make your cut-flower crop different from your competitor's. The chances are that all growers will produce a white sweet pea, whatever named variety they opt for. But they won't all produce a glorious crop of peonies, which are so space-greedy and have such a short flowering season. So choose well – grow what you love and what works best for you.

There are many kinds of scabious, all beloved of wildlife. This variety is 'Pink Mist'.

Perennials are the plants that will make your cut-flower garden different from your competitor's down the road. It's likely that everybody farming their garden for flowers will grow sweet peas, cosmos, ammi, larkspur – a classic mix of cottage-garden annuals in easy-to-use colours for weddings and kitchen tables alike. While your varieties of sweet pea may be different from your neighbour's, the chances are you'll both be growing a white, a pale pink, a pale blue (for weddings) and something more striking (the dark red 'Beaujolais', for example, or the blinding salmony pink 'Daily Mail' for a bit of fun and for your market flowers). Where you *will* differ from your neighbour, though, is in your choice of perennials.

Floristry is only gardening on a much smaller scale, and, like gardeners, florists are always looking for interesting and original combinations, colour mixes and texture choices, which might make them different from their competitor. Your perennials will bring you customers, whether you're selling to a wholesaler or to local florists, or are doing your own floristry. Through a long season of dahlias there is a huge variety of perennials that come and go, keeping your dahlia bouquets fresh; presenting new challenges to the florist in you to make the plants you have work hard for you.

But the reason you won't be growing all the same perennials as your competitor is because of the space they take up. Annuals are a quick-turnaround crop, and while they might be worth less per stem because of the sheer quantity of them, the flower farmer will make more from them because he or she can grow, crop, sell, grow, crop, sell. . . constantly making use of the same piece of land. Perennials are a different matter, and you need to think carefully about the space you might devote to them. Those that fulfil a double role – make good foliage as well as cut flowers; provide a windbreak as well as give you material to cut – are twice as valuable as those that flower for just a few weeks once a year.

How to grow perennials for cutting

When people start out commercializing their gardens, they often find it hard to get away from the idea that they're making a garden. Habit drives you to dot your plants about in a mixed border so that it'll look attractive. This is also a great way to fight disease, because any kind of monoculture can attract disaster, which, if you're gardening organically, is hard to fight until the garden is established.

However, for ease of cutting, let me recommend planting in rows or blocks. Unless you're planting acres of asters, you'll still be growing in a way that nobody would consider a monoculture. For the same reason you'd plant lettuces in a row –

A bumbler making a beeline for the eye of a delphinium and 'lunch'.

speed of cutting – you should also plant your delphiniums all together.

Cutting flowers from an established herbaceous border, where the different plants are spread out among each other, will slow you down. Think of the time you have available, and the number of flowers you plan to cut each day, and that should inform your planting plan. Five bouquets a week out of a beautiful, classically planted herbaceous border is one thing; but fifty bouquets a week and you'll soon get fed up with walking up and down and up and down to pick just three stems each of whatever it is you're looking for from that clever and attractive planting scheme. And you must remember that you're planting for bouquet making, which is miniature gardening: it is when you come to put the bouquet together that the attractive repeat patterns of colour and texture can happen, in the same way they do in a border.

If you want to make gorgeous herbaceous borders, be a gardener. If you want to be a flower farmer, think like a market gardener, not an RHS-Gold-Medal-winning garden designer.

At Common Farm we move our perennials about quite ruthlessly: perennial beds can quickly get choked by bullying plants and will really benefit from all the plants being dug out, split and moved on every two or three years. Again, if you were establishing a great big gorgeous herbaceous border you certainly wouldn't dig it all up every year. But you're thinking like a market gardener, remember, and so your perennials need to be working for you: by being split, by being turned into more plants, by being thrown away if they're under-performing or if you find you're not using them. And remember also that perennials being used for cut flowers rather than a cloud of colour in a garden need feeding too. At the end of the season, mulch them; throughout the season, feed them. Be kind to them and they'll reward you with lovely crops.

Growing from seed

By all means beg cuttings of perennials when you're starting out. Take cuttings of your neighbour's lovely lavender in late summer and you'll have the beginnings of a free hedge next spring. And take advantage of bargain plants for sale: the cheapest at plant sales are always alchemilla, because it seeds itself so freely and so nobody wants to buy it. I always buy *all* the alchemilla plants I see, because we'll need about 200-plus stems for an average June wedding, and we can't have too much of it. (It never has a chance to seed itself anywhere at Common Farm because we cut pretty much every stem.) But if you're looking for a good, reliable section of delphiniums or perennial asters, or echinacea, then you may as well grow a tray of it yourself, as you need to plant

Plant perennials in mid-June at the same time as your biennials, and the young plants, like these echinacea, will be ready to plant out in September for flowering the following year.

Learn to recognize self-seeded plants in your patch – this is a small white mallow, which I find very useful for wedding flowers in July.

The Chelsea Chop

As a person commercializing your flower garden, you might think hard about the 'Chelsea Chop': the time around the date of the Chelsea Flower Show, about 20 May, when, traditionally, gardeners cut a third to a half of the height of strongly shooting perennials. You certainly increase the number of flowering stems by cutting off the top third of a flowering plant at that stage, but the *height* of those stems will be shorter if you chop, and so those chopped plants might not be so useful to you later in the season. You may find that you're quietly chopping quite a lot as you go anyway: we have a beautiful black-leafed loosestrife cultivar ('Firecracker'), which reacts very well to a Chelsea Chop in a traditional garden, especially if the front of the patch is chopped and the back is left to flower earlier and give a graded height to the plant. We cut that lovely black foliage as filler from the minute it starts to shoot in April, and so it gets a Chelsea Chop of sorts, if a rather unorthodox one, by being cut continually until it dies back in October. The poor plant never gets to flower for us, because the foliage is more valuable than its delicate spikes of tiny yellow flowers.

Some plants in fact react very well to a Chelsea Chop for the person farming their garden: heleniums are one example, as unchopped they can be very leggy and not give you much in the way of flowers.

By Chelsea Chopping you can delay flowering as well – so if you have a large clump of, say, perennial asters, chop half of them and you'll double the length of the flowering season of that patch.

If you want to make gorgeous herbaceous borders, be a gardener. If you want to be a flower farmer, think like a market gardener, not an RHS-Gold-Medal-winning garden designer.

strategically for your market. And don't forget to make the most of your plants that do self-seed – learn to spot them among the weeds!

So order your perennial seed with your biennial seed in March. If the instructions on the packet recommend a period of cold, you can always pop the seed into the freezer for a few weeks to aid germination. Sow the seed along with your biennials in early June, make sure the trays don't dry out as the summer heats up, keep an eye on your seedlings, and do prick them out and pot them on. This way, you'll have some nice-sized seedlings to plant out in September, when you're clearing beds and have some space ready for them. With a fair wind and a gentle autumn, the plants will have established themselves well enough to survive the winter, so they can flower for the first time the next season.

Buying plants

Established perennials can be expensive to buy. Probably the best thing to do is to make yourself a wish list of perennials for your cut-flower garden and keep it to hand. Some will be easy enough to grow: sea hollies, asters, delphiniums, echinacea are all easy to grow from seed. Others you can pick up at plant sales for very little (namely the alchemilla I mentioned earlier). And even for those plants that don't come at such a bargain, plant sales are still brilliant sources of stock for the start-up flower farmer. You will find unusual varieties, often wonderfully strong plants, and a fount of knowledge from the people selling them. They may not be raised organically, and in that respect

might not be compatible with your approach, but if you're going to farm your garden organically then I suggest that you do so only 'from the gate in'. The people selling at plant sales are often incredibly knowledgeable horticulturalists – so if they use peat-based compost and the plants are dotted with the odd blue slug pellet, the pellets are easy enough to brush off into the bin before you plant out, and you can take the view that *you'll* live by example and garden peat-free.

Save 10 per cent of your income as your first year goes by, and at the end of the year look at what you have in your account. You might decide that now's the time to invest in a more serious collection of perennials. You'll have had a season to see what sold well and what grew well. You'll have seen which perennials you find easy to grow from seed, and you'll have your first patch of home-grown delphiniums bushing up nicely for an impressive early crop next year.

So now is a good time to think about where you'd like to spend your savings. Call your local plant nurseries and ask how much you need to spend to get a wholesale price. Look at the rose growers and perhaps plan a patch of roses for cutting. Always ask how many of anything you need to buy before you get a better price than the advertised retail price. You may not feel comfortable asking for money off or pushing for good prices, but it's worth it! You might get as much as 50 per cent off if you buy enough, and that's a great many more plants to your budget. Of course there'll be moments when you just see and buy: after all, you're a gardener, and no gardener can resist a plant nursery happened across down a country lane. But think of the gorgeous rare plants you buy there as personal treats: they *may* end up being brilliant croppers for your cut-flower patch – but plan to spend strategically at the end of the year and you will get more for your money.

Propagating from established plants

You should propagate from established plants, not only to make more plants but also to keep the original plants looking fresh and performing well. Perennials can get clogged up with roots and old growth in the middle, and while they might seed themselves about the place, and make lots of lovely plantlets at their edges, the old centres can do with a really regular chop.

Think seasonally: spring-flowering perennials like to be split in the autumn; autumn-flowering ones like to be split in the spring. Of course, time constraints may dictate that you do it all at the same time – we tend to do most of our lifting, splitting and rearranging in the autumn, no matter when the plants are going to flower, just because it fits our schedule more effectively. From January onwards we are planting great quantities of seed and potting up, potting on and planting out, and so in the autumn we try to go through the whole garden rearranging everything. Gardening comes with all sorts of rules, but

you'll do best to bend those rules to suit your schedule, or you risk ending up not doing the jobs at all.

There are many ways to split a perennial, but for us the basic tool is a sharp knife. You can lift plants and jam two garden forks back to back into the plant and pull the plant apart, but I have *never* been able to do this very effectively. The sorts of plants that come apart this way seem to dig up easily in chunks, and so that's what we do: dig up a chunk at a time and make new plants this way. We take a carving knife to plants that won't come apart easily with the back-to-back fork method. It's the same knife we use on the dahlia tubers in spring. Don't sue me if you cut yourself: be careful!

Which perennials to grow?

Perennials, as I've said, are space-greedy – and they stay in the ground, using that space whether performing or not, all year round. You might crop a useful perennial for only six weeks a year. And before cropping, you've had to propagate it, grow it on and wait, in some cases for three or four years, until the plant is mature enough to produce a useful crop.

I'm not saying you shouldn't grow perennials. As explained earlier, your perennial choices are what will make your garden unique, and a good collection of successionally flowering perennials will raise the game of your cut-flower-patch team, giving your crop an edge against the competition.

But when choosing your perennials, think hard about the length of their performance season as well as their potential performance in the vase. Is it worth growing those heavenly peonies when their flowering season is only a fortnight long, and that fortnight almost impossible to predict? Plant half an acre of peonies and you still can't promise your bridal clients which ones will be in flower on a given date. If you adore peonies and think that unique, delicate peppery scent is worth the space the plants will take up for just that fortnight of flowering, then you should grow them – after all, the foliage on a peony is also very useful, especially late in the season when it turns a gorgeous dark red colour. If you have good, well-drained soil, plenty of mulch and worship their gorgeous blowsy flower heads, then by all means grow peonies. If you're growing for yourself and have the space, then I would certainly recommend three or four peony plants. What I'm saying is that, if your garden is to work for you as a commercial venture, these are questions you need to ask about *all* your perennials.

There are, *I* think, better – if more workhorse – performers with which you might fill the space demanded by those peonies. I always tell cut-flower growers that they can't have too much alchemilla edging their borders and paths. "But it'll seed itself everywhere," protests the experienced gardener. "Not if you're cutting it for weddings it won't," I tell them.

Think about the classics useful for weddings: astrantia in all its different colours is a winner. Roses, so long as you grow the kind you can't find in ordinary high-street florists, will work well for you. Don't try to compete with the African and Venezuelan growers where roses are concerned – you can't beat their price, but you certainly can win on scent and many-budded sprays of gorgeousness (see Chapter 7). So long as you cut them back

Sea hollies are easy to grow from seed, and are popular for weddings, where they're useful because they stand well out of water.

judiciously after their first June/July flowering, many roses will give you a worthwhile second flush in September. Sedums can be cut in the green, through flowering, and dried for Christmas wreaths; ditto sea hollies and globe thistles. Echinacea's petals bruise easily when used in floristry, but the buds make lovely buttons in a bouquet, and the thistleish centre of the flower, without the petals, looks amazing mixed in with dahlias in September.

Recommended perennials for cutting

The list of perennials you could grow for cut flowers is inexhaustible. So again I say grow what you love, experiment, and think outside the traditional floristry patch to make your perennial collection interesting to your potential customer.

I will, however, give you a list of my favourite perennials (not including roses, which get their own dedicated chapter) for cut flowers.

Achillea, in all its glorious colour variations, makes a fine short-lived perennial.

Aquilegia, granny's bonnet, columbine. . . Whatever you call them, these flowers signal to me that spring is really sprung. Cut them from mid-April to late May.

Achillea

From the frothy foliage in the early season to the waving heads of colour later on, achillea is a great favourite of mine. There are so many lovely colours to grow of this gorgeous flat-headed umbellifer. However, watch out for aphid infestation: I grow achillea in part as a sacrificial plant near my sweet peas, because I love my sweet peas even more.

Aquilegia

When these delicate flowers open I know that spring has arrived. They will seed themselves prolifically about the place, and the challenge to stop them interbreeding is beyond me, so mine are all a mix of blues and purples and pale pinks which they've created themselves. I'm not complaining.

Alchemilla

Alchemilla ('lady's mantle') should be grown especially by those wanting to do their own floristry. Throughout early summer, it will be the most useful

WHAT WE'VE LEARNED

If cutting or working with achillea or alchemilla in bright sunshine, you might find your skin reacts to it: by the end of the summer I always have odd marks up my arms from working with alchemilla. The marks fade over winter, but return the minute I start cutting it again in the spring. Protect your skin with gloves and long sleeves.

I say grow what you love, experiment, and think outside the traditional floristry patch.

filler in your garden, taking the place of the early-flowering spurges as an acid-green light in your arrangements. After a first flowering, if you cut it back hard, it will flower again in September in time to join the second flush of roses – a mini-spring at the season's end. And don't forget the foliage: useful for buttonholes, or to collar a bouquet. (Never look at a plant in the cut-flower garden and ask only what its flowers can do for you.)

Artemisia

This is wonderful for its silver foliage as well as to keep the moths out of your drawers. I'm not mad about the flowers, but the foliage is indispensible: we use artemisia throughout the season for silver highlights. It can be a bully in a border, and will propagate itself by layering at will.

Astrantia

These pretty, bobbing, crown-shaped flowers are incredibly useful for weddings, as they don't wilt easily. Grow 'Hadspen Blood' for a lovely dark red variety, or the classic white *Astrantia major* for weddings. Astrantia has lovely foliage for framing bouquets too. It will do well in a fairly shady spot.

Bleeding hearts

With a similar shape and texture to Solomon's seal (see page 81), bleeding hearts have gorgeous pink flowers instead of white. Both are recommended, however, if you have space, though their flowering season is short (so, again, think how much you love them or will use them if you don't have much space).

Campanula

Tall campanulas, or Canterbury bells, make good cut flowers, though in my experience they're not enormously long-lasting in the vase. They work as short-lived perennials for me, though you can grow them as biennials too. Grow them for weddings or other special events. They will get rust if they're too wet, so if you have a dry bed, or better-drained soil than our clay, they will be more successful for you than they are for us.

Cow parsley

If you're lucky enough to have it growing where you live, the wild cow parsley (see Chapter 11, page 176) is incredibly useful: you can't have too much cow parsley in the early season, before your ammi starts to flower. If you don't have it growing wild, or if you want a bit more interest, you could invest in the black-stemmed 'Ravenswing' variety, much beloved of garden designers at the Chelsea Flower Show. It's very attractive in a white border as a dark, purplish understorey, and it would be worth a little extra if you're selling by the stem to florists or a wholesaler. However, you will be cutting most of the foliage off, and the leaves won't get much of a look-in in a bouquet. I save my money and don't bother with the cultivar – but then I've got cow parsley at hand in the fields.

Delphiniums

These classic cottage-garden flowers offer tall spires of white, blue and pink in early summer. We don't grow a huge number of them because we grow a lot of their more delicate cousins, the larkspurs, instead. Their leading flowers can be enormous and very heavy, which is useful for big event floristry, but not so much for everyday bouquets. They will keep flowering on more delicate side shoots though, and if cut back and mulched will produce a useful second crop in September.

Echinacea

Good to cut when the flowers are just forming, when the petals are just showing. Later, when the petals are fully out, echinacea flowers can be easily bruised and bashed in the floristry process, though I will admit I strip those petals and use the gorgeous bristly central cone of the echinacea flower as wonderful architectural structure in my bouquets.

Euphorbias

In a great many forms, from the annuals *Euphorbia marginata* and *E. oblongata* to the wonderful perennial *E. dulcis* 'Chameleon', with its gorgeous red foliage, euphorbias give all year round in the garden. Watch for skin reactions to the caustic sap when cutting it, however. You should condition euphorbias separately from other flowers, as the milky sap will turn water cloudy. It's best to cut it to the length you're going to use it, so that you don't have to cut the stem for a second time and cause the water to go cloudy again.

We use the woody stemmed perennials throughout winter for greenery and foliage: they're structural, have wonderful developing flower heads, and are a favourite haunt for overwintering ladybirds (a staunch ally of any gardener – see Chapter 11).

Geraniums

We do grow hardy geraniums for cutting. They have fragile flower heads, so should be cut early, before the flowers open, but their shape and structure in a bouquet add air and light to an arrangement.

Sun spurge, a lime-green euphorbia flowering in April and May, will self-seed into cracks and edges, as here in front of this bed. It makes a wonderful filler, though beware the caustic sap on your skin.

Grow geums for buttons of highlight colour – home-grown floristry is like creating miniature gardens, and a tiny splash of contrast can bring a bouquet to life.

Geums

I'm fond of the cheerful pop of a geum in a flower arrangement; as a cropping flower, though, I'm not sure the yield is worth it in a small space. If you have a little space in an ordinary border, then you might want to include a few and cut them for floristry. The same job is done at the same time of year by strawberry flowers, which for the same space are more abundant, though perhaps a little short-stemmed.

Giant scabious

This big lemon-yellow scabious is an absolute favourite at Common Farm Flowers, but the plant itself takes up so much space that I may yet grub it out and throw it away, despite my love of its flowers. In a good season it will flower from June till October – but in a dry season it'll be over in a fortnight. The base of the plant is space-greedy and it self-seeds freely, the little plants sending down such deep taproots that even small seedlings are difficult to grub up. This is a classic example

of a plant that doesn't give enough back to the flower farmer in comparison to the space it uses – and an equally classic example of a plant that we grow here because it is so loved. You may think me mad, but gardening for cut-flower production, as with any gardening, is so much driven by personal taste: this is why your garden will work differently from mine, and why competitors are competing on a taste basis as well as on a price/product level.

Hellebores

These shy winter and spring flowers are better cut than you might think. So long as you get them direct into water and keep them there, they will last for up to a week. Support their freckled faces with other foliage in an arrangement, so that the customer can see their beauty. They will interbreed ruthlessly, so if you're planning a marvellous collection of different colours you'll have to be careful to control marauding self-seeded hellebore children.

The black-leafed loosestrife 'Firecracker' and tellima – they are both bullies and like similar damp, shady conditions, so will work well together.

Heleniums

These can bruise easily as cut flowers, and even with good conditioning sometimes look a bit wilty, but with the benefit of a good Chelsea Chop should give you a nice mix of interesting late-season flowers. You might find room for a few, but in a restricted space I think there are better performers.

Hostas

If you can keep the slugs off it, hosta foliage is marvellously structural as a collar for bouquets, and I think it very striking in arrangements. Perhaps choose an interesting bicoloured variety, or a sharp, acid-green-leafed variety – if you're going to grow perennials for cutting, then always choose striking colours and shapes, which other growers may not have. There are enough different kinds of hosta that all the growers in the country could grow a different one and so keep their spring bouquets unique.

For the battle with slugs, it's worth trying the chilli and garlic dip that can be used to protect bulbs from mice (see Chapter 5, page 86). It's helpful against slugs too, though needs frequent reapplication in wet weather.

Loosestrife

We grow a big patch of the black-leafed loosestrife 'Firecracker', for foliage rather than its teeny yellow flowers. It can be wilty, but if well conditioned it works well with support in a mix. It can also be invasive, but we plant it with other bullies such as the wildflowers tellima and bistort, and let them fight it out between themselves.

Oriental poppies

Prolific self-seeders, these make surprisingly useful cut flowers. They are difficult to condition when the flowers are out, but cut them when the flowers are beginning to break through the armour-like carapace of their bud heads, when the tissue-

Monkshood warning

Monkshood, a member of the buttercup family, is very poisonous and can kill if ingested. The poison can also be absorbed through broken skin. I'm always nervous when I see it in arrangements, and we don't grow it at Common Farm. Your flowers are inevitably going to be touched by your customers, if only when they're throwing them away. So I say be wary of this very poisonous flower, however beautiful, and take care if you're going to grow it.

When oriental poppies open in a vase, they 'explode' as they burst out of their carapace – great fun for children.

paper petals are just showing – like the silk fabric of a Tudor gentleman's slashed sleeve – and sear the stem for 30 seconds in an inch of boiling water. Without removing the stems from the boiling water, fill up the container with cold water and let the flowers condition. The petals will pop out, exploding that hard bud casing in a most satisfying way, and flower in the vase as though they were in the ground.

If for no other reason, grow them because they make unusual, if fleeting, cut flowers and are hard for your customers to get hold of. Their seedheads are also lovely in autumn decorations and Christmas wreaths. Any flower that does more than one job for the cut-flower grower is worth its place in the garden, and poppies try hard enough for me.

Ox-eye daisies

All the different cultivars of the traditional wild field daisy are useful as cut flowers. From the classic wild daisy to the lovely shaggy heads of *Leucanthemum* x *superbum* 'Osiris Neige', these

daisies will be very useful in the garden, flowering from early June until the first frosts. Don't use too many together in a bouquet or a scheme, as they do make the water stink.

Penstemons

These make perhaps unexpectedly good cut flowers. I find the lightness of their spears of bell flowers a charming and delicate contrast to a heavier spike like a delphinium, and the fact that they flower well into the autumn makes them late-summer-useful too.

Persicarias

Persicarias flower in various forms (including wild): throughout the spring with the *bistorta* variety, and into the autumn with the 'JS Caliente'.

The tall spikes of the flowers make them lovely airy additions to bouquets. They can be bullies in a border and will spread, so perhaps plant each in its own edged patch to prevent invasion.

Phlox

The glorious scent of late-season phlox in the garden is one of my favourites. Phlox is a classic cut flower, and its large blooms are valuable because of their size. They do the same job in a bouquet as sweet rocket and lilac do earlier in the season, and together with the panicle hydrangeas take flower farmers through the late summer into autumn. If you can't grow hydrangeas because your soil won't suit them, then do think of growing phlox: it's an easy cut flower to find a market for, and you will certainly find colours to suit the themes you're growing.

Rosebay willowherb

Rosebay willowherb makes an unusual cut flower, though some of you may be sucking your teeth in horror at my suggestion that you grow it to crop. But look at the white cultivar, so graceful a spike of height in the border. It won't seed itself everywhere, because you'll be cutting it continually. Think like a cut-flower grower, not a traditional gardener, and your opinion of many of the herbaceous bully gang might change dramatically.

Scabious

We grow a good variety of scabious, both annual and perennial. Their beautiful flowers make cheerful buttons in a bouquet, cut very well, and are perhaps a bit stronger-stemmed and more prolific than cornflowers, which do a similar job in floristry. Cut their seed-heads too, for attractive structural shapes.

Sea hollies

Anything so multipurpose as a sea holly gets my vote in the cut-flower garden. They are good for cutting from the first green buds (wonderful buttonhole material) to the crunchy dried heads, which sparkle in a Christmas wreath. If, like us, you garden on wet clay, be sure to add a lot of grit when planting out your sea hollies: they won't thrive in a bog.

Sedums

Sedums are useful as a cut flower, from the first silvery budding of the flower heads to the great flat plates humming with wildlife when the flowers come out in September. Grow lots of different colours, because your bees will love you for it as much as your customers do.

Freshly split sedum clumps produce finer stems than established clumps, and smaller flower heads. You might need some big-headed, long-stemmed sedum for the wedding and event market, or some smaller-headed, finer-stemmed flowers for posies and bouquets. If you have a sedum collection, then split half of it one year and half the next, to provide you with both kinds of stems.

Solomon's seal

If you have a shady spring garden, then solomon's seal is a great frond of loveliness to sell in the spring. When I include it in bouquets people seem to be amazed that it stands so well in water. For me it will last ten days plus when cut, and the graceful curved stem of leaves with little white bells hanging off them make a light foil to the heavier stems of tulips and iris, which flower at the same time.

Verbascum

A prolific self-seeder of tall spikes of flowers, verbascum offers colours to suit every palette. We grow a lot of the rich, dark variety 'Purple temptress' to complement pale early-summer colours. For us, verbascum does a spiked job in a bouquet, which is taken over by the penstemons when they begin to flower a little later in the summer.

Verbena bonariensis

Although we grow this as a perennial, I know that lots of people find it difficult to make it overwinter. If this is the case where you live, treat it as a half-hardy annual (see Chapter 2, page 51). It's a slow starter to grow from seed, but you can take heeled cuttings from a friend's plants and they'll root on nicely. The little cushions of bright purple flowers bobbing about your garden on their tall stems are as beloved of butterflies as of florists, though for floristry you should cut them when the flowers are still young, or they're likely to shed themselves everywhere and not do their work in your bouquets.

Veronicas

There are over 500 different kinds of veronica, so you can pick and choose to suit your gardening taste. They grow, flower for a short time, divide easily – and make very useful little spikes in flower arrangements. Although their flowering season is short, I think they're very good value in a cut-flower garden.

We grow verbena bonariensis as a perennial here. Leave its dried stems to protect new growth over winter.

Five top tips for growing perennials for cutting

❀ **Choose carefully:** you can make your product very different from your competitor's by growing unusual perennials.

❀ **Grow perennials from seed in June.** This is an inexpensive way to make sure your garden has good, sturdy, new flowering plants for the following season.

❀ **Think twice before you Chelsea Chop:** do you need those long stems you'll be sacrificing by chopping back your perennials in June, or would you rather have more flowers on shorter stems?

❀ **Grow what you love.** Perennials are space-greedy: don't give space to a plant that doesn't thrill you when in flower.

❀ **Split regularly, and feed throughout the season for best results:** perennials need loving if they're to perform to their utmost.

Chapter five

Bulbs & corms

Bulbs are expensive, and there will almost inevitably be cheaper ways for you to supply tulips, still relatively locally grown, to your customer. But I'm not sure I know a single flower farmer who sticks to their bulb budget. A carpet of narcissi or tulips in spring is a swathe of hope when there's still frost on the ground. But you need to spend wisely for a crop that does more than cover its costs.

Despite the expense, I challenge you not to plant tulips in your cut-flower patch.

Growing bulbs in a cut-flower patch seems to be a given, however economically senseless. Not only do the Dutch grow cut flowers from bulbs for a fraction of your cost, but there are also still big growers of bulbs in the UK – who will undercut you almost to the rate of selling each of their cut-flower stems for the price you'd pay per bulb. There's little you can do to compete with this, unless you plan to join their number and plant bulbs by the hundred-thousand yourself.

Though you and I may be flower farmers, we are gardeners too, and however often I might chide you to remember that your patch should be laid out with ruthless economics in mind, I'm not sure I know a single grower who doesn't put in at least a couple of hundred tulips every year, let alone a sack or two of narcissi. They convince themselves that these bulbs will be left in the ground

to be harvested year after year, and then, inevitably, in the rush to cut for market or bouquets, they speed-cut the leaves as well as the flowering stems, leaving nothing for the bulb to absorb back into itself for use in producing flowers next year; or they lift and chuck the bulbs to replace them with other flowering plants in order that their bulb patch keeps producing year-round rather than for just a month or so in the spring.

So, despite the financial burden, I think you'd be hard pushed to find a cut-flower patch without a bulb section, despite the fact that a bulb-and-corm budget tends to outstrip the modest annual spend of £200 on seed, £200 on dahlias, £200 on long-lasting, high-performing roses. . . Bulbs are not only planted for less return per bulb than for other flowering plants, but are also often used as annuals and thrown away after a season. (Note: I have included corms in this chapter, as it is usual to order bulbs and corms from the same suppliers, and their treatment is often the same.)

If you're growing a garden from which you will cut only *some* flowers, then you will get more out of your bulb investment than the ruthless, cut-'em-all flower farmer, as those that are left uncut will be

WHAT WE'VE LEARNED

Plant bulbs because you love them, treat your bulb budget as an indulgence, and remember that you can order bulb flowers in quite small quantities from wholesalers if you want tulips and narcissi more cheaply than you can grow them.

Posies of spring bulbs sell like hot cakes at farmers' markets.

able to absorb through their leaves all the goodness they need to flower next year.

Remember, though, that if you plan to leave your bulbs in the ground between flowering, then you'll have to work around them and over them in between times – being careful not to dig them up by mistake or spear them murderously with a border fork when working the bed for another crop on top of the bulbs. You must also take care to leave the leaves when you cut – which will shorten the available stem length and therefore the potential price you'll get per stem, but may also mean you get a second flush of a smaller, more delicate flower, which is truly lovely in a tulip. . .

You'll also need to replace some of your bulbs year-on-year, as you'll inevitably cut some too hard and they won't re-shoot. The other difficulty is that fashions for flower colours change as often as fashions for party dresses, so an investment in a thousand ballerina tulips might look fantastic in your bouquets this year, but next year you might long for creams and whites.

I have given in to the inevitability of a bold bulb budget (ours is currently about £600 per annum – next season I fear it will bust the £1,000 mark), but am getting better at not spending money where I don't get enough return. I experiment each year, but close-kept notes remind me where *not* to spend.

How to grow bulbs for cut flowers

Growing bulbs for cut flowers requires you to step away from what you may have learned over the years about planting bulbs four times their depth, with plenty of space between them so they can bulk up into clumps and spread. Growing flowers for cutting is about speed and turnaround. You plant, the flowers grow, you cut, you then quickly use the space for something else.

You'll need lots of drainage, but not necessarily very deep beds, a strong back for bursts of quick planting, and careful planning – so that you use the bulb space effectively before, during and after the bulbs' tenure in that particular patch of ground.

Beware: bulbs are an attractive food source for some animals, notably mice, but also squirrels and birds. Protect them from marauding mice by either dipping in hot chilli and garlic (see box overleaf) before planting, planting in a wire net, or putting annoying (to the mouse) short sticks poking up out of the ground where the bulbs are. We do none of the above, and lose about 10 per cent of our crop to the mouse.

Buying bulbs

Check for height before ordering: there are tall varieties of grape hyacinths, for example, but you must look for them. Grape hyacinths are bred

As soon as your bulb order arrives, take the bulbs out of the boxes to keep the air circulating and prevent rot.

short for planting at the edge of the border, and, while any of them are lovely cut in a posy, a short-flowering stem will reduce your potential profits and the versatility of this gorgeous scented spring flower, much beloved of early-awaking butterflies.

Choose scented varieties: you are competing with big growers who will be selling much more cheaply than you, so spend a little extra on ordering especially scented and unusual varieties – you can use these to bulk out the cheaper flowers you might have sourced from the bigger growers.

Tell your bulb supplier exactly when you want your order delivered. Order spring-flowering bulbs in early August, to avoid finding that the special varieties you particularly hanker for are sold out, but ask your bulb supplier to deliver as close as possible to the time you intend to plant. If you split your deliveries into daffs/narcissi, alliums and then tulips, you'll end up paying for three deliveries. We order all our bulbs to be delivered towards the end of September, when our wedding season is slowing down and we can be confident that we

can get the daffs/narcissi in quickly, and that the alliums and tulips won't have time to rot if we're only storing them for a few weeks until it's time to plant them. If you do need to store your bulbs, for even a short time, take them out of their boxes to keep them well aired.

Don't forget that you need to order bulbs in spring too. Bulb-ordering is a job that you might think is done in August, when you've finally made your agonizing choices and pressed 'send' on an order that is bigger than you'd originally budgeted for. But summer- and autumn-flowering bulbs should be ordered early in the year. I order my gladioli in February or March for April planting; ditto my acidanthera. Late-season bulbs are a surprising and delicious addition to the autumn garden, and ignoring their potential would be to the detriment of your late-season cut-flower choices.

Planting bulbs

First choose whether you want to naturalize your bulbs or whether you're going to use them as annuals. If you're naturalizing them in grass or beds, they'll need space to bulk up, they'll need to be planted much more deeply, and when cutting you'll have to remember to leave the leaves alone, to absorb goodness into the bulb for next season's flower production.

If you're using your bulbs as annuals, then they can be planted much more shallowly (saving your back but not your purse strings), and so close together that they're almost touching. Your tulip crop, for example, might be finished the first week in May, when you can quickly lift the bulbs and compost them, top-dress the bed they were in with a little compost, rake the surface down to a fine tilth, and direct-sow a nice crop of tender, late-season annuals in the space.

You might find that you plant the more expensive bulbs – alliums, say – to settle into a bed and come back year after year, but grow the cheaper ones, such as tulips, as annuals. Another reason for these different approaches is that you'll probably always want the same alliums, as their colours are limited to blue, purple and white, but your taste in tulip colour may change year after year.

Bulbs carry all the food they need for flowering in the bulb itself. All they need for successful growing in the ground is good, well-drained soil and a little water. Make sure you give your bulbs plenty of drainage, or you risk them just rotting away and disappearing altogether. Dig in grit if your ground tends to get waterlogged during the winter.

Accepted practice is that you should soak bulbs for an hour or so before planting them. Our ground is boggy, and soaking the thousands of bulbs we plant would slow down the process too much, so (except with anemones and ranunculus) we don't. You may have fewer to plant, and the time and energy to soak them – and you may, as a result, get a better crop than we do.

For naturalizing bulbs, plant at four times as deep as the bulb is large. If using bulbs as annuals, just twist them in to well-cultivated ground so that the growing tip can be covered. (Though remember that a shallow-planted bulb may prove an easy draw for your neighbourhood mouse community.) The difference between the two approaches is summarized in the table below.

Bulbs for annual cut-flower production	Bulbs for naturalizing
Plant fairly shallowly.	Plant at a depth of four times the bulb's size.
Plant in hundreds, close together but not quite touching.	Plant in groups with space between so that they can bulk up over time.
Cut the flower stem right down into the bulb, for length of stem.	Be careful not to cut the leaves, which the bulb will need to draw goodness down into it for next year's flowering.
Plant in beds, for ease of lifting after flowering.	Plant in grassy areas or in beds. Flowering bulbs In grassy areas are beautiful in a garden, but bulbs which have to compete with grass won't be as strong-stemmed or large-headed as those grown without competition in beds.
To make the best use of the soil, direct-sow green manure over bulbs that will be lifted, then dig the green manure into the soil as the bulbs are removed.	Direct-sow hardy annuals over bulbs that will be left in the ground after flowering.

For spring-flowering bulbs, the planting schedule in autumn is as follows:
September: daffs/narcissi
October: alliums
November: tulips.

For summer- and autumn-flowering bulbs, the schedule isn't so tight. We tend to plant them in April because our hardy annual and biennial crops should keep us busy until the end of July. We like our gladioli later, to go with our dahlias.

Bulbs for spring flowering

If you do decide to invest in bulbs for a spring crop, then your season will start earlier than a competitor who saves their pennies and relies on their hardy annuals to begin their season – though bulbs do flower according to temperature. This year, for example, we've had a very warm spring here, so the spring-flowering bulbs have flowered six weeks earlier than they did last year, when we had a sudden cold snap for several weeks in March.

Recommended spring-flowering bulbs

Take time to trawl through the bulb catalogues. Plant colours that suit your garden's palette. I'm sure that some of you will disagree with my recommendations, but, as ever, never plant what you think you ought to, but plant what you love: you will find it much easier to sell your product if the colours are those that sing to you.

Alliums

Don't be put off by the inevitable oniony smell as you cut them: it disappears after an hour or two, and alliums make marvellous cut flowers. From the extremely inexpensive ornamental onion *Allium cowanii* to the easy-to-grow-from-seed garlic chives; from lovely purple chives to that classic purple sensation, the July-flowering drumstick onion *Allium sphaerocephalon*, alliums in fact rank among my favourites.

Don't grow the very big heads unless you want to use them for huge arrangements at weddings or events, or plan to dry them for Christmas, as they invariably squash in a bouquet and lose their stunning shape. *Allium sphaerocephalon* will naturalize in long grass if you'd like it to, which can be space-saving, though remember that if plants are competing with grass you won't get such a prolific crop as you would if they were in a bed.

A good stock of spring bulbs will keep you going until your hardy annuals begin to flower.

Allium sphaerocephalon, or the drumstick allium.

Anemones are a great early crop to grow in your polytunnel – their smiley happy faces enormously cheering when it is still really quite wintery outside.

Anemones

Anemones are lovely early-blooming flowers (from February) when planted in a tunnel. They can be short-stemmed and are easily weather-bashed, but at a time of year when choice is limited, a crop of these exquisite, friendly-faced flowers is enough to cheer the most humbugging customer. We use them from February to May for floristry, weddings and bouquets. They are the forgotten jewel in the British-grown cut-flower calendar.

When you've prepared the space you're going to plant them in, soak the corms for an hour to encourage them to start shooting. We've found that doing this with anemones transforms the crop quantity. Although we don't soak our several thousand tulips/daffodils/narcissi, nor our alliums, our anemones are planted under cover, and the soil around them is drier than the soil outside. Tradition has it that cut anemones prefer to be kept in shallow water, but we've not experienced any particular fussiness in this respect, and we condition ours in the same way as we do any other cut flower: two-thirds up to the neck in water.

Brodiaea

Brodiaea sports gorgeous blue, pointed daisy-shaped flowers from mid-May in the mildest areas; later elsewhere. You could fleece half your crop to bring them on earlier, in order to stagger your flowering season.

Camassia

The wild, original version of this plant is the quamash: a squat, bluebell-blue spike of a flower native to North American prairies. Camassia, its cultivars, come in varying intensities of blue and white. This is a great plant to naturalize in long grass, which is what we do at Common Farm. The bulbs are too expensive to buy new each year, so we have them in our meadow spaces, where they flower early in the year. Cut them young, when the first flowers are opening low on the stem, and they'll last a good week in a vase. Don't arrange them to stick out of bouquets horizontally and expect them to stay that way: camassia always reach for upright and will turn themselves up at right angles, which can look odd.

Daffodils and narcissi

Daffodils are members of the narcissi family. I mention both names here because they're generally split as such in bulb catalogues. Those known as narcissi are better known for their scent than those known as daffodils. However, do not dismiss the cheery daffodil when looking for spring bulbs to buy: there are many interesting varieties which might give your spring crop a zing. I love the pink trumpet daffodils, which are exquisite with blue scilla in posies, as well as the pure whites. Unusual daffodil varieties can be expensive to buy, so you may choose a few each year to naturalize and then perhaps buy a sack of cheaper 'Early Cheerfulness', for example, to bulk out your crop.

Order your daffs and narcissi in August, and *think* before you order just because the sack is cheap! You'll be competing with your Cornish and Scilly-Isles colleagues for daff and narcissi sales: can you beat them on price? Very unlikely. So don't buy daffs or narcissi for value; buy them because the varieties you choose are different, and will not look like the 10-for-£1 bunches in buckets outside every florist in the land. You are growing for sale – so consider your market and give your customers something different.

I'm told that the big Cornish and Scilly Isles growers put their bulbs in one autumn, and leave them to flower the next spring, crop them the spring after that, leave them to flower the following year, and then crop them one more time before replacing the bulbs the following year. Do try this yourselves – and let us know how you get on!

Freesias

Grow freesias under cover, and watch out for marauding squirrels. The flowers are delicate, a beautiful round bell shape with several blooms on each stem, and their rich, peppery scent is amazing . . . but I wonder if it's worth growing something that is imported in such quantities and so cheaply?

Hyacinths

At the time of writing, there is no large-scale grower of hyacinths for cut flowers in the UK that I know of. So there's an opportunity for somebody! I would certainly buy hyacinths in quantity every January and February if a UK grower took up the challenge of producing them for market.

Irises

These make great cut flowers. Cut them when the buds are just beginning to split and show the flower colour. Enjoy choosing your colours. We leave our irises in the ground to increase in number year-on-year, and so take care when cutting to leave enough leaf for the corms to reabsorb for next year's crop. For us, irises are highlights, not fillers.

Be aware of the difference between the June-flowering bearded irises (*Iris germanica*) and the May/June-flowering Dutch irises (*Iris x hollandica*): the former don't last so long in bouquets but make amazing flowers for weddings and events; the latter are a little smaller and last longer when cut.

Wild (foetid) irises (*Iris foetidissima*) are of a beauty and delicacy unparalleled in the garden. Their flowers are great for weddings, but don't last more than a few days in a vase. I don't find the scent unpleasant, but would be wary of using a great many of them in a scheme, for fear others might.

Lily of the valley

In the olden days (not long ago) there was any number of lily-of-the-valley growers in the UK. Now these short spikes of deliciously scented, perfect miniature bells can be difficult to get hold of. So grow some and you will have a market for them. They don't travel well, so your local flower shop might well snatch them out of your hand. If you sell them on a market stall they'll be gone within minutes. And there isn't a bride in the land who will refuse lily of the valley for her bouquet.

The plants like dry shade and without competition will proliferate happily. If a friend has a patch, you can dig up a chunk of it once it's flowered and plant it in poor, stony soil. To force it to flower early, pot it up and bring it indoors when it's just poking through the soil. Or grow it for your first season indoors then put it outside to naturalize.

Ranunculus

Ranunculus come in so many gorgeous colours, not necessarily easy for the high-street florist to get hold of, that you can really indulge yourself – investing in white with pink picotee edging, and so on. For brides who want English country flowers, and don't know what they can have instead of roses or peonies in April, offer ranunculus: they are often won over by them at first sight.

Grow them in a tunnel for length of stem and protection from the weather. Watch out for botrytis and mildew: water well and feed, feed, feed to help them withstand attack.

Ranunculus will start flowering from the end of February in a warm, frost-free tunnel.

Snake's head fritillary

You're more likely to have seen this exquisite, chequerboard-headed spring bellflower wild in swathes over ancient meadows, and so it may surprise you to discover that the bulbs are relatively inexpensive and that they make great cut flowers. They're rarely grown to cut, and I doubt would travel well out of water, and so will have rarity value as a crop to sell locally – whether to your local florist, at market or in bouquets of your own design. Worth a try. See Chapter 11 (page 179) for more details.

Snowdrops

These can be ordered in clumps 'in the green', i.e. freshly dug up just after they've flowered, to plant out in early spring. They're not by any means showy flowers for big bouquets, and some superstitious folk would frown to see them brought into the house. I believe very strongly in absolving all flowers from undeserved connections, and cut snowdrops at will in season! They have a

Snake's head fritillary will establish itself happily in a meadow (so long as the mice and birds don't eat all the bulbs) and makes a great cut flower.

honey scent, which on a dull day in February will fill the air around them with sweetness.

Snowflakes are similarly shaped, stronger-stemmed bulbs. They flower slightly later in the season and make good, if short-lived, cut flowers.

Snowdrops in a shallow glass vase. Some people are superstitious about bringing snowdrops into the house – not I!

Tulips

Tulips get an extra-long entry in this section, as they're likely to represent quite a big part of your bulb spend – and because they're such favourites of mine. The points I make here about permanent underplanting, buying cheaper elsewhere, and so on, might equally be applied to other bulbs, but I've chosen to make them in relation to tulips.

When ordering, check for height: the stem length doesn't have to be 50cm (1'8"), but less than 30cm (12") will be less use to you. If you're planning to naturalize your tulips in a bed, then it's difficult to cut a shorter-stemmed tulip in such a way that

the foliage is left undamaged to help feed next year's flower bud. In my experience, the later the tulip flowers, the longer its stem. However, a tulip flowering in mid- to late May is a great deal less use than one flowering in April, when tulips can be the main crop in your cut-flower garden. By May, people have had enough of tulips, and they compete unsuccessfully with peonies, the first roses and sweet peas.

Plant tulips late to avoid disease – November is the best month, as a good burst of cold weather before they start shooting helps kill off diseases. Tulip fire will burn the leaves of your tulips and turn the flowers into twisted, odd-looking specimens

A spring posy of apricot tulips, pheasant-eye narcissi, cowslips and spring greenery, with blossom provided by spiraea 'Bridal Wreath'.

The Viridiflora tulip 'Spring Green' – not one you'll find at most high-street florists, and so worth growing yourself.

Make the most of your space

If you are planning to leave your dahlias planted in your beds rather than lifting them to overwinter under cover, then you might think about planting tulips through and around them – after all, if you're farming your garden for profit you may as well maximize space, and, if you can, minimize time. This is a Sarah Raven tip for getting two crops from the same place (you could try it with roses too). This system of cropping will work well if you're not going to cut your tulips too hard.

Alternatively, if you're lifting your dahlias, you might consider putting your tulips in the space they occupied. We rotate our crops this way: dahlias then tulips, then, once the tulips are cropped, we lift and compost the bulbs and direct-sow a crop of late-season tender annuals in the vacated space, so our beds are worked effectively all year round.

no good for sale. In a mild winter, tulip fire is a risk that all growers fear. I've been talking to our bulb supplier in the past week, and she's concerned because this year we've not had any cold weather to kill off disease in tulip bulbs.

Tulips cut with still-tightly-closed buds, packed flat in boxes out of water and stored in the cool, will keep for a week if you need to hold them back for a special order. When you take them out of storage, condition them by snipping the stems to re-open the drinking cellulose cells, and stand the flowers in clean, fresh water.

To keep tulip stems straight rather than curling about as they grow in the vase, you can condition them in cones of newspaper or prick their stems just below the flower with a pin. I prefer the curly-wurly look myself, and so never take either of these tulip-controlling precautions.

Again, choose unusual varieties: they're more expensive, but you can't compete on price with the big growers for ordinary tulips. People will be coming to you for something unusual, and five baroque-looking parrot tulips go much further in a bouquet than twenty ordinary tulips.

Bulbs for summer and autumn flowering

Plant gladioli and crocosmia for summer flowers, and acidanthera, gladioli, nerine lilies, crocosmia (again) and schizostylis in spring for an autumn show. The summer- and autumn-flowering bulbs and corms can really transform what might otherwise be quite an ordinary-looking crop.

Recommended summer/autumn-flowering bulbs

The following is, of course, a list of my favourites. Each has a special job to do in our garden, and I wouldn't be without any of them. You may have others to add – or may dislike some I've chosen. And, as always, you should think imaginatively about what you can grow for cut flowers, consider spaces where you might grow for cropping as well as for beautiful borders, and don't overlook the more unusual plants to crop in addition to the more obvious.

Acidanthera

For late-summer flowers this bulb cannot be beat. Ours entirely failed last year: there were plenty of healthy leaves but not a single flower, and we weren't the only people to see not one flower on their acidanthera. The previous summer had been hideous – almost sunshine-free and a total washout. We think this was the problem: that because the previous summer was so bad, the bulbs weren't fully fed. We will buy again next spring, as a couple of hundred acidanthera bulbs will fit in ten largeish pots and give you a glorious crop of delicate, late-summer bounty: bobbing heads to dance above your dahlia bouquets. You can certainly plant your acidanthera in the ground if you have a warm bed with good drainage. We grow ours in pots because our ground is often boggy and we have little space. If creating a border as well as a cut-flower patch, potted-up acidanthera can be plunged in to the border as they flower to fill a late-summer gap.

Agapanthus

This is another one for a hot, dry place. We grow them in pots in the yard, as our soil is too boggy, and even the teensy crop we get a year, of maybe a hundred flowers, always gets used up instantly. With a warm, well-drained environment you could have a whole bed of them for July/August.

Acidanthera is exquisite, and so richly scented that you should use only a few in a bouquet, or the smell can be overwhelming.

Growing bulbous plants can be a labour of love – these agapanthus seedlings may flower in their third year – but the rewards are worth the effort.

Agapanthus are expensive to buy as bulbs, and space- and heat-greedy to grow in winter – but irresistible. This one's called 'Apple Blossom'.

Crocosmia

Crocosmia is a good dual-purpose cut flower: the arching stems of blinding colour at the end of the summer are useful in bouquets and sell well as individual stems, but if left uncut the flowers make striking curved fronds of boot-button-sized seed-heads, useful for autumn wreaths and Christmas arrangements. Crocosmia's colours and shapes make a good foil for dahlias. It's also the sort of plant you find dug up in chunks for sale at local fairs and can be picked up relatively cheaply. We grow some naturalized in long grass and some in beds – this way they flower at different times and at different strengths, giving me two different products to play with.

Gladioli

The glorious shocking-pink spires of species gladiolus flowers can be seen in season swathing the roadsides in Cornwall, so much do they like

the conditions there. Here at Common Farm they disappear a season after a measly flowering of about half the number we put in. We battle on to make a hot enough bed, with grit spaded in by the barrowload, but have yet to win in the gladioli stakes. But these delicate, smaller-headed glads are too much a temptation for us to ignore, especially where weddings are concerned, and so we labour on.

For colour, style and sheer amusement value, on the other hand, the traditional 'Dame Edna' gladiolus cannot be beat. Stunning in wedding flowers, valuable per stem, coming in every colour of the

Growing wild in Cornish hedgerows (it could even claim to be a wildflower) the species gladiolus makes a great cut flower. Sandy, well-drained soil suits it best.

rainbow, these big, brash cultivars will make you laugh, if nothing else, as they blind you with their brilliance in serried ranks of good jokes.

Their corms are inexpensive: order them in spring with your acidanthera, but do wait until the soil has warmed up before planting them in beds, or alternatively pot them up and keep them under cover until the soil has warmed up. So long as you have good enough drainage and you stake them well, your crop should be gratifying, It's also worth planting successionally, to keep your stock coming, though with luck they'll settle in and work as a perennial if you want them to.

Gladiolus nanus, the more delicate glad, closer in appearance to the species variety, is gorgeous in a hand-tie. Grow the white for brides.

Cut gladioli when the bottom flowers are just beginning to show colour. They should stand a week in water easily.

Nerine lilies

In a hot, dry spot these rich-coloured, very-late-season South African lilies will settle nicely and bulk up to be a good annual crop. They are incredible in late-season bouquets – exotic-looking, long-flowering, and come in some great colours as well as the usual blinding pink (though I do love the blinding pink too). Whites, salmons, lemony orange-pink. . . If you have the right environment for them they make a great flower to crop.

Ornithogalum

In dry, light soils this almost allium-looking waxy white flower will naturalize happily for you. It makes a good, unusual and therefore valuable cut flower as well as a striking addition to a dry border.

Schizostylis

Another late-autumn-flowering bulb of South African origins, despite its provenance, schizostylis doesn't mind heavy soil or wet conditions. With tall spikes of flowers, it comes in varieties ranging from pale pink to dark red.

Bulbs for the Christmas market

When you order bulbs in August or September you could order enough to pot up some to force for Christmas gifts. A handful of paperwhite narcissi in a galvanized bucket shooting through a little grit and moss makes a great present for somebody to buy at a farmers' market. Amaryllis, hyacinths, crocuses – all these pot up well for Christmas presents.

You won't be the only person potting up bulbs for Christmas, of course, so make yourself a little different by keeping an eye out for vintage cups and saucers, teapots and soup tureens through the year at second-hand shops, brocantes or flea markets – you might find you have a nice collection to plant up before the festive season.

A word on the law

By all means plant paperwhite narcissi, hyacinths and amaryllis in pots to sell as growing plants, but beware the medium you grow them in. Don't add any kind of manure to the compost, as there are strict rules relating to the risk of diseases carried in manure (however herbivorous the creator of it).

Bulbs for forcing can be planted shallowly, with their tips standing proud of the surface of the container they're being planted in. You can plant them shoulder to shoulder, really crowding the container. Use very gritty compost for good drainage (especially if using quirky containers without drainage holes). They will flower fairly quickly inside, and you should warn your customers not to over-water them, so they don't drown.

If you receive your bulbs in September and forget to put them somewhere cool, they will begin to shoot strongly before you plant them in mid-November, and will flower before Christmas, so do remember to store them really cool (and mouse-free) until you're ready to bring them into the warm to force them. Keep them cool enough that they won't shoot: in a fridge, a cold cellar or a cold barn. If you keep them in the bags they arrive in, then do check for mould from time to time.

Second-hand fridges are fairly cheap to buy, and a fridge is enormously useful as a store for seed, for sown seed trays in which the seed likes a bit of cold to encourage it to germinate, and also for bulbs before forcing. If you can find one, a large cheap fridge with temperature control might be very useful.

Amaryllis

Amaryllis like some real warmth to get them going, so you can always give them a boost with a heated propagating mat, or you might find they're a bit sluggish for Christmas.

They are expensive bulbs to buy and, without heat, difficult to grow quickly. They are most beautiful, though, and last for weeks in the vase. You might invest in a few bulbs and just grow some as an experiment. I will admit that although

Amaryllis will flower two or three times from the same bulb over a season. Here, a final fling mixed in April with pear blossom and grape hyacinth.

Hyacinths make wonderful cut flowers if you can get them to give you long enough stems. These were forced indoors for a bouquet in January.

I've grown them to cut for years, this year I have given them up: not only are they expensive to buy, but they're also too space- and heat-greedy for the results. I need a giant, heated polytunnel to make a successful amaryllis crop.

Hyacinths

Plant hyacinths in trays of half grit and half compost and put them somewhere cold for 12 weeks before bringing them indoors in batches to flower successionally. They'll be barely budding in cold storage, but as soon as you bring them in they'll shoot to flower in six weeks. Put them somewhere they have to reach for the light, to encourage better stem length: under a dining-room table or under a potting bench in a warm greenhouse are good places.

WHAT WE'VE LEARNED

We're now in year four of our cut-flower business, and I'm seriously considering finding the money to invest in a second-hand refrigeration unit – the sort that are used in lorries. So long as I can adjust the temperature, it'll be incredibly useful as a chiller of seed and bulbs in the winter, but also to keep wedding flowers cool in the summer, when they're done but not ready to be delivered.

Paperwhite narcissi

The term 'paperwhite narcissus' is strictly translated as *Narcissus papyraceus*, but it is generally used loosely to mean the small, many-headed, highly scented narcissi commonly forced for Christmas and the winter season.

Paperwhites will flower in as little as six weeks from the date of planting – plant them outside in September and they'll be flowering in November.

So if you're potting them up for Christmas, work backwards and plant them two weeks into November. If you arrange for your bulb order to arrive at the end of September and plant a couple of hundred outside too, they'll make a useful autumn catch crop if you have space for them. Their stems outside won't be enormously long, but for posies, table centres and winter wedding bouquets, they're very useful. Our particular favourites here are 'Ziva' (strictly a paperwhite) and 'Grand Soleil D'or' (not strictly a paperwhite).

Five top tips for growing bulbs for cut flowers

❋ Consider whether you are going to naturalize your bulbs or use them as annuals and plant accordingly. To naturalize, plant them four times the depth of the size of the bulb and be careful not to cut the foliage when cutting flowers. If treating them as annuals, then you can plant them more shallowly and don't need to worry about saving foliage when cutting.

❋ Protect your bulbs from marauding mice, unless you are prepared to lose some of your crop to them.

❋ Remember that there are bulbs to have for flowering year-round, so plan your planting accordingly. They do not only make a spring crop.

❋ Use a permanent bulb crop to underplant dahlias and roses, usefully doubling up the space.

❋ If adding grit to increase drainage in a bed for bulb planting, don't add it in a layer at the bottom, as this simply creates a sort of sump or riverbed for water to sit in. Mix grit into the bed instead.

A few words from Karen Lynes
of Peter Nyssen Flower Bulbs & Plants

Karen Lynes is a fine example of the sort of supplier a flower farmer needs: one who always has time to answer an email, make a recommendation, or find that elusive variety you're looking for.

" We provide around 3 million bulbs for cut flowers to the trade and to private customers. Home-grown flowers are picked at their best and freshest – and you just can't buy that quality in the chain shops. Tulips, narcissi and hyacinths are in the top category for cut flowers: from December until mid-May you have a steady supply of cut blooms, and the fragrance that most give is a welcome lift on dull winter days. Anemone De Caen varieties, with their array of colours, have also become more popular.

Alliums are good for extending the spring season, and you can also spray the seedheads to use in Christmas decorations. Ranunculus have a delicate rose-like appearance, but are tough. Snowflakes are perfect for April-to-early-May weddings. Amaryllis are excellent cut flowers: buy large bulbs (at least 30cm/12") and you will get at least two stems from each bulb, with 3-5 flowers per stem. Early-flowering gladioli, like the Nanus varieties, provide the first flowers of summer from June.

Naturalized bulbs need care to encourage them to reappear each year. Always feed with a high-potash fertilizer and deadhead regularly. Most bulbs need moisture to flower the following year. Tulips are often treated as annuals, though some varieties, such as the Darwin Hybrids, will flower for a few years. Always deadhead bulbs before the flower starts to fade, to stop seed production at the expense of the following year's flower.

If you want to force bulbs to flower early, they need a spell in the dark and cold – usually at a temperature between 1°C and 9°C (34-48°F), no more – to fool them into thinking they have gone through winter. As a rough guide, tulips and smaller varieties of narcissi need 15-20 weeks; grape hyacinths 12-15 weeks; non-prepared hyacinths 6-8 weeks. ('Prepared' bulbs have been chilled by the supplier before delivering.) Give all bulbs around 4 weeks in warmer conditions (15-20°C / 59-68°F) once they start sprouting.

Prepared hyacinths, and narcissi paperwhites and 'Grand Soleil D'Or' are best for Christmas forcing. Prepared hyacinths need 10-12 weeks in the dark below 4-5°C (39-41°F). When they are about 6cm (2") high, bring them into cool light until just before flowering. Put paperwhites and 'Grand Soleil D'Or' in the cool dark for 2-3 weeks to set the roots, then bring into the light. "

Chapter six
Shrubs

Shrubs are not what you might immediately think of when planting a cut-flower patch. But without them, where will your greenery come from? And how will you protect your delicate flowering annuals from the ravages of a summer storm? If only we had planted more shrubs when we started! Choose the right shrubs and they'll work as hard for you in your flower garden as any dahlia.

Guelder rose and wild viburnum, or wayfarer's tree, in a mix with a variegated dogwood and lots of flowering garden plants for a big display at a wedding.

Shrubs work so hard in a cut-flower garden that I'm surprised at how sidelined they can be in a plan. If we had known how our business would develop, shrubs would have made a much greater dent in our earlier budgets. Especially if you plan to do floristry yourself, then you should certainly consider early investment in shrubs.

Shrubby plants provide structure within which to frame your flowers, strength to support flowers that need a bit of scaffolding, and great windbreaks in your garden. In a confined space, try to avoid plants that only do one job for you: nigella gives you flowers and seedheads – photinia gives you lovely red leaves and a windbreak. When we started, we planted a great deal of native hedging, which, while good for wildlife, etc. and certainly ticking our 'eco' boxes, left us without good shrubs for cutting. We're making up for this

now, but we are playing a game of catch-up, and, although shrubs are expensive, we've been forced to buy larger specimens rather than propagating them ourselves – which we could have done easily when we started out and were less busy.

How to grow shrubs for cutting

Remember, shrubs are as variously fussy as any garden plant: shove them in to fight long grass and they'll fight instead of growing. Put them in a bed with plenty of space and they'll thank you kindly and flourish. And I urge you to leave your shrubs for at least a year before giving in to temptation and cutting them, even for a teensy bit of foliage. Give them time to settle in and put on new growth: that way they'll do *much* better and

I appeal to you to put aside any horticultural snobbery that may have infected your opinion of some garden shrubs.

not die on you instantly when you do start to cut from them, the year *after* you planted them.

Which shrubs to grow?

Think about what you're going to need your shrubby plants for. Are you going to supply your customers only during the summer months? In that case, choose lots of interesting deciduous shrubs with vibrant summer colour. Are you planning to supply for Christmas? In which case, pick evergreens for winter foliage, as well as some of those heavenly scented winter-flowering shrubs. For example, Christmas box is a must for suppliers thinking of fresh green Christmas wreaths.

If you're looking for windbreak value as well as foliage from your shrubs, then some evergreens will help, though of course the wind-thrashed side of a hedge risks suffering from windburn in a vicious winter, and won't make good cut material. And finally – before you make your choices – I appeal to you to put aside any horticultural snobbery that may have infected your opinion of some

garden shrubs! My mother instilled in me, very successfully, a dread of laurel, and a super-dread of spotted, striped or any variegated version of this trusty garden hedging plant. True, once you've got it, it's very hard to be rid of – but it's hardy; its leaves are lush at Christmas; and for big wedding arrangements it gives a wonderful dark understorey against which your roses, peonies, larkspur and sweet peas can glow. . .

You may have an inherited horror of variegated shrubbery, considering it tainted by the fact that supermarket carparks plant great blocks of it. Stop! Think! Why are supermarket landscape architects so fond of these shrubs? Because they look good even if they're mildly neglected, they look good when cut back hard, they make effective and attractive hedges quickly – working hard in an environment where the point is to sell milk, bread and salmon steaks. These garden designers have chosen wisely, and you might even do worse than go down to your local supermarket and see what they've chosen. The shrubs that do well locally for others will do well also for you.

Of course, we all have our own taste in plants – which is why we're gardeners. At the other end of the spectrum from the car-park shrubs, if you like, native shrubs have their own place on the stage. For details of these, see Chapter 11.

Recommended shrubs for cutting

The following are my top shrubs for cut-flower use. We cut from our gardens all year long and so we grow for both winter and summer use. I know

---

WHAT WE'VE LEARNED

I was always intent on having lots of flowers to put in my bouquets, when I might have spent a little more time worrying about what I was going to *frame* those flowers with as my business grew.

there are some florists out there who proudly state that their bouquets aren't filled with foliage, but, for me, foliage is as important in floristry as flowers: not only for the structure it gives but also the texture, the developing buds in spring, the bouncing bunches of berries in the autumn. Our bouquets are like miniature gardens: foliage and greenery stops a bunch having a simple annuals-in-the-cutting-patch look and transforms it into good-herbaceous-border. We cut every stem we use in our floristry, so wouldn't waste a minute cutting something that didn't have an important role to play in a bouquet. We cut a lot of greenery.

Bay

Slow-growing but invaluable in winter arrangements and for Christmas wreaths, bay will keep bad witches away from the front door of the house, or so I'm told. I put it in bride's bouquets too. The buttons of developing bay flowers are very pretty in arrangements, as are good shiny leaves. Grow your bay out of the way of the wind, because it's easily weather-worn and then no good for use with cut flowers.

Buddleja

We find the white buddleja especially useful in floristry, and cut it a great deal throughout August and September, when the side shoots will keep flowering long after the main first flush of flowers has been cut. The foliage is pretty too, through the autumn, when silver fronds frame dahlias beautifully. Of course you'll have to fight the butterflies off the buddleja when cutting it, but if your cut-flower patch looks anything like mine, they'll have plenty of other forage to find.

Choisya

Some people hate the smell, but I find that the scent disappears an hour or so after cutting. Choisya is a very useful foliage plant, and the white flowers in spring are a beautiful harbinger of summer. It doesn't wilt and it works well in foam-based arrangements as well as in water.

Christmas box

The honey scent of Christmas box is a highlight of the winter. Bring a spray into the house and your whole house will be filled with scent from the tiny little white flowers which climb the branches of this diminutive shrub. Tuck a sprig into your Christmas wreaths, or add a little (the scent is strong – too much in a bouquet might be overpowering) to January or Valentine's Day bouquets (it flowers throughout the winter).

Corkscrew willow and hazel

For use in Mothering Sunday and spring bouquets, the corkscrew willow and hazel, with their catkins growing from charming curly-wurly stems, are incredibly useful. We pollard our corkscrew willow quite ruthlessly, which means we get lots of lovely fresh thin stems for use year-on-year.

Elder

This is sometimes a difficult one to condition, but as foliage in the early spring and autumn (you won't need it so much in June, July and August, it's most wilty months) it is very useful. I've never managed to make the flowers stand in water, but the berries are beautiful in autumn arrangements – or frozen to be used in game gravy during winter – and I find myself cutting the wide, generous pairs of leaves a great deal, especially in autumn.

Elder will send up fresh useful shoots if cut hard back in the winter. You don't want thick, woody stems for arranging with: the new shoots from a well-pruned elder will be easier to condition than woody stems, and the leaves will not wilt so easily if the drinking cells of the cellulose in the stems are young and fresh. Like smoke tree (see page 110), I wouldn't use elder in flower-foam-based arrangements, but in fresh water it should last at least five days. It's extremely easy to propagate – you can layer it or take cuttings. Look for cultivars with interesting leaf shapes and colours: the one I use is dark red.

Euonymus

Variegated euonymus is an extremely handy shrub for floristry from the garden. I love the white-and-green in winter because it has a frosty-edged look – very useful in wreaths and garlanding. Whether variegated or not, euonymus is evergreen, it doesn't wilt, it's smart in buttonholes and, though slow-growing, I'd say it's a pretty serious contender for the top five most useful shrubs for cut-flower production.

Honeysuckle

Don't forget that in a confined space you always have room for climbers, whether up the side of buildings and fences or over pergolas. Just because something's climbing doesn't mean it can't be useful as cut-flower material. Honeysuckle is a delicious addition to a flower crop – and the

Don't forget to grow climbers for cutting. Honeysuckle makes a wonderful cut flower, but don't use too much, as the scent can be stinky.

same goes for clematis. I use honeysuckle a lot, and if you're short of space it is certainly worth considering. Be wary of using too much summer-flowering honeysuckle in a scheme, however, because it can go from making a breathtaking scent to causing a stink in no time.

The winter-flowering honeysuckle does what it says on the label ('*fragrantissima*'). Winter colour (as in big, showy flowers) is rare in plants, but scent, in tiny flowers on this and others, such as winter-flowering viburnums and Christmas box, is easy to find, so bring cut branches indoors to scent a house with the promise of spring.

Hornbeam

Hornbeam is especially useful for bouquet floristry in the spring, when its leaves uncurl so attractively on the stem, but it's also handy in the summer, as its wide sprays of leaves make good frames for bigger arrangements. It can tend to wilt when cut, so do sear it to be sure of success (see advice for lilac, below, and also Chapter 12, page 189).

Hydrangea

If you have the ground for them, then grow hydrangeas. Their glorious flower heads are high on my list of favourites for wedding flowers. You can cut up the big mop heads and use parts of them in floristry, you can dry the flowers for Christmas, and you can so often use the foliage too.

As a rule, hydrangeas need free-draining, moist soil in part shade to really flourish. Here at Common Farm we have no free-draining soil and so have built a raised bed for our hydrangeas, in the shade of a hawthorn hedge. We'll see how they do there. But until they're established, as a florist I'm in the market for good-quality hydrangea heads from as soon as they start to flower, at the end of July, right through until they're frost-bitten in the autumn – so if you *do* set yourself up as a British hydrangea-cropping flower farmer, let me know and I may buy from you!

Lilac

Some people are superstitious about lilac in the house, but I say absolve this deliciously scented and opulently flowering shrub of all associated bad luck and cut it ruthlessly when it's in flower. (Do check with brides, though, in case their grandmothers have strong feelings on the subject.) Lilac can be very fussy to cut, so it needs searing when you cut it: strip the foliage, split the stem with a sharp snip to a length of 2-3cm (1"), and dunk the cut end of the stem straight into 2 inches of boiling water for 30 seconds. The container can then be filled up with cold water so that the flowers can condition as normal.

You can keep your lilacs growing in pots. Plant the pots into beds for most of the year, and then a

Some people are superstitious about having lilac in the house, but I bring it in by the armful.

few months before you'd expect them to flower, bring them into a warmer environment like a polytunnel, forcing them to flower earlier. When they've finished flowering you can 'plant' them out in their pots again. Equally, lilac clippings or prunings taken in the winter can be brought into the house to force. Lilac is easy to propagate from cuttings and will send out runners, which you can dig up and pot up to make more plants if you need them.

Mahonia

I'm a fan of the foliage of the smaller varieties: there's a dark-red-leafed (almost black) variety which is fantastic in Christmas decorations and winter bouquets. Admittedly, in a scheme with limited space this might not be high on your list, but for unusual shape and texture it's certainly worth considering. Mahonia will not only feed your bees through the winter but will give you an interesting variation on the usual varieties of foliage available from cut-flower growers.

Myrtle

Myrtle offers white flowers in July as well as very useful winter foliage. It is tender, so grow it protected against a warm wall. Myrtle is associated with weddings, because it is traditionally the emblem of love and marriage, so it's a very useful shrub if you intend to do wedding flowers. A sprig of myrtle is great in a buttonhole and makes fine, delicate greenery with which to set off flowers in a bouquet. There are variegated varieties that I like too, which, especially if cut in winter, will give a frosty-edged look to an arrangement.

Philadelphus

If you're planning to grow for big-wedding floristry, then the mock orange is just the plant. The scent is breathtaking, and an established philadelphus won't mind being hacked at for floristry purposes too much. If it's in full flower when you cut it, it will stand better in water if you sear it before conditioning. Cut to the length you're going to use in your floristry, and sear it as you cut it (see details opposite, for lilac, and also page 189).

Physocarpus

In all its many colour variations this is probably my favourite shrub. It's deciduous, so not much use in the winter months, but from the first sign of budding leaves until the frost-singeing winds of early November, I cut my physocarpus all summer long. Like smoke tree, plant your physocarpus in three sections: one for spring cutting, one for high summer and one for autumn. The yellow varieties are my favourite in springtime, but the dark-red-leafed kinds are great with dark-coloured dahlias late in the season. They have very pretty buttons of tightly packed tiny flowers in June, which are useful to cut for bouquets when freshly out, but also make great seedheads for Christmas decorations. Physocarpus is easy to propagate from

Physocarpus 'Dart's Gold': if I could choose only one shrub to grow for my cutting garden, it would be this one.

cuttings, so buy a few established plants to start yourself off, and take it from there.

Pittosporum

For winter use, pittosporum is brilliant. Plant the variegated, or the dark-red variety 'Tom Thumb', which, though a dwarf shrub, will grow plenty big enough to cut from. Pittosporum is reportedly tender, but we find that so long as it's not in a dank place, it thrives almost anywhere. In a dank environment it sulks. We keep some in the polytunnel and some outside.

Skimmia

Although tender, in the right place skimmia is a very useful winter foliage/flower plant. You may wonder why you bother with it throughout most of the year, but in deep winter, when you're really stuck, it comes in extremely handy.

Smoke tree

This shrub can be difficult to condition, and I'd be wary of using smoke tree in arrangements using flower-foam-based floristry. But cut directly into water, and conditioned overnight, it works very well for us as a striking foliage plant. Plant it in three sections in your garden and use them separately in spring, summer and autumn: this way you'll keep your stock in good condition and have nice long stems for when you need them. It propagates easily by layering and also via cuttings, and has some really fantastic different-coloured cultivars. Deciduous, it's no use in winter. Cut back hard in the autumn to encourage long, straight, fresh growth in the spring.

Spiraea

So many different varieties to choose from! For leaf *and* flower I love the spiraea cultivars. From 'Bridal Wreath', with its elegant curving sprays of tiny white flowers, to *Spiraea douglasii* for its incredibly useful, fine-stemmed foliage (I'm not so fond of this variety's doll's-house-sized loo-brushy flowers, and tend to cut them off before using the foliage). This is an easy shrub to propagate. I find it most useful in spring – and as a low windbreak.

Viburnum

Viburnum opulus 'Roseum' (the snowball bush) is a classic in the cut-flower-look book. It flowers for

Grow spiraea 'Goldflame' for gorgeous foliage throughout the season – turning burnt red in the autumn – as well as for pretty cushions of pink flowers in spring.

The snowball bush is a great shrub for cut-flower material.

together are dramatic, interesting, and difficult to get hold of unless your customer is prepared to get to Covent Garden very early in the morning to snap up the little that's on offer there in season. Viburnum makes great hedging, and the leaves are attractively edged with red in the autumn.

Don't plant too much too close together if you are in an area where viburnum beetle will ravage the leaves and render your crop useless. We split our viburnums up by mixing them in with our native hedging and, so far, don't suffer viburnum-beetle attacks. Our neighbours have ripped up their viburnum hedge rather than witness the results of beetle damage, but our plants seem to be protected by their hawthorn and wild dogwood neighbours, and so we keep on cutting them as a crop.

Winter-flowering viburnums, such as viburnum bodnantense, make lovely lacy-flowering fillers as well as good greenery.

just a fortnight in May(ish – depending on your spring), but, like lilac, is exciting enough (I think) to grow just for that fortnight's glory. Especially if you grow for weddings, snowball bush and lilac

A late spring allows for a bride's bouquet with early roses, late anemones and white lilac – a rare combination as a result of the weather.

fruit blossom pruned in January. In a warmer house this should happen more quickly.

A great many spring-flowering shrubs and other plants will also force happily if clippings are brought indoors, so do the same with, for example, clematis, willow, forsythia and lilac. Experiment with different shrubs as you prune in your garden, to see which ones work well for you.

Using tree blossom

If you have any fruit trees, or even a bit of an orchard, then you have another great source for floristry material. When you prune your fruit trees, put your prunings into a bucket with a little water and bring them into the house to force them. With enough prunings, you can force them succession-ally: bring the first bucket into the house and leave the second outside or in a barn or shed.

If you have a great many prunings – enough to have stock week by week over several weeks, or even a month or two – then heel them, cut end, into the soil somewhere outdoors and then bring them in as and when you want to force them. In our cold house (kitchen about 17°C / 63°F on a mild day in winter), it takes about a month to force

Five top tips for shrub planting and use

❋ If you're planning to do floristry from your cut-flower patch, grow shrubs. You will need them for foliage as much as you need flowers.

❋ Choose shrubs for their flowers and berries too – making them work hard in the cut-flower garden.

❋ For summer use, choose for interesting leaves and colours; for winter use, choose for scent.

❋ Take care when cutting and conditioning shrubby material with woody stems. You may need to sear the stems for 30 seconds in boiling water before filling the container with cold water for a good few hours' conditioning time.

❋ Use shrubs to make windbreaks and so save space.

A few words from Stephen Read of Reads Nursery

Stephen Read runs Reads Nursery, a specialist fruit-tree nursery in Suffolk. Fruit-tree blossom, and often fruit itself, make a useful addition to floristry, and both branches of blossom and sprays of fruit are ingredients that your customers may not be able to find easily and so might be happy to buy from you.

When planning your cut-flower patch, think hard about where you'll source your growing material from. A good plant nursery will give you incomparably more in information, support and advice than your average chain garden centre, and the specialists you will find in such places are often generous with their knowledge and advice.

"Using tree blossom, particularly that of fruit trees, is a useful way of gaining extra value from a tree.

Blossom grown in the field, either in the form of a productive 'hedge' or free-standing tree, will be somewhat erratic in its timing each year, as it is so weather-dependent, but there will be a regularity to it in most seasons. Varieties are grouped by flowering time, so choose carefully and you can have a spread of blossom that lasts several weeks.

Certain fruit trees blossom well and reliably, such as apples (crab or dessert types are both very good) and also pears, which have a simpler blossom colour and structure. These are also fruit trees that do not mind regular pruning, which is an important consideration, as pruning wounds, if indiscriminate, can lead to disease and the eventual downfall of the tree. Using a systematic approach, it is possible to harvest the majority of blossom without long-term harm to the tree.

Stone fruits, such as plums, cherries and gages, are generally not so useful for harvesting blossom. Pruning these trees early in the year will often leave open wounds, which are risky for these particular trees, as they can lead to infection by silver leaf disease.

Consider your local climate, as a few other choices of tree are available if a sheltered spot and space allows. Apricots will fruit on regular spurs, and, while the blossom is early, the trees can often be pruned with care to provide a reliable supply of pure white blossom (and later some fruits if not all the blossom is taken). Peaches, if grown under glass, can provide a splendid display of pink blossom in early March.

Catkins from hazels and cobnuts are also plentiful, giving floristry potential in early season from February."

Chapter seven

Roses

How often have you bought a bunch of imported roses, sniffed them, shrugged and accepted that they have no scent? As an artisan flower grower you can't compete with the big importers for stem length or endlessly repeating identical flower heads, but you can win on scent. Grow roses for their gorgeous perfume and their abundance.

When customers come into our flower studio and it is full of roses cut for a wedding, they stop short, and I see them breathing in the glorious scent of what we grow. Nobody should have to accept scentless roses: a rose with no scent is like unfinished silk, a painting with no colour, a sunny day clouded over. It is a sadness, a disappointment.

And what shrubby perennial plant grows almost as well as anything else here in the UK? What shrubby perennial loves our gentle climate and our cool summers? What flowers from early June until late September – and even, with endless deadheading, will still be flowering on Christmas Day? Why, the English rose! Of course, not all of you reading this will have an English country

garden at your fingertips: you may be in Seattle, Sydney, Syracuse – or Somerset, South Africa – but the generic rose available to you in your local florist shop will almost certainly be long, straight-stemmed, single headed, and flown in from South America or Kenya. And forgive me for a certain partiality, but when I say 'gorgeous garden rose', do you imagine a serried rank of identical roses, or the dream of an English country garden, where the roses festoon arches and frame seats, and scent the air with a heady mix of apricot and lemon, honey, ginger and a shake of cinnamon? My guess is that you see the latter.

So, wherever you're growing cut flowers, for your own use or for sale, do consider whether you can

'A Shropshire Lad' in a late-summer bouquet with grasses.

Nobody should have to accept scentless roses: a rose with no scent is like unfinished silk, a painting with no colour, a sunny day clouded over.

grow roses. A fresh-cut garden rose, full of life, its stamens just visible through a coy, curling skirt of taffeta petals, seems barely related at all to its poor cousin, hydroponically, chemically, factory-grown in a giant tunnel on the side of a Venezuelan mountain.

A bucket of scented, home-grown roses for sale by the stem at a farmers' market, in your local farm shop, at your nearby home-interiors shop or vintage-style specialist in gardenalia, will likely sell out as you deliver them. Absolutely perfect, stripped of their thorns, with not a hint of black spot on the leaves, perhaps each with a little brown card label attached on which a note can be written – these roses might sell for as much as £2.50 each; in London twice that. And you're wondering whether you can make some money growing cut flowers?

How to grow roses for cut flowers

There's a slightly different approach when grow-ing roses for cut-flower production as opposed to in a mixed border. David Austin recommends that roses planted for cut flowers be planted much more closely together than they would usually be: they don't need to be more than 30cm (12") apart. The idea is that the crowding of them will encour-age long (and therefore valuable) stems. Roses in a mixed border usually have to compete for air and light with underplanting and with other shrubs, but in a bed for cut-flower production they should be weed-free, not fighting for air with dense under-planting (and therefore needing mould control), and so can be planted more closely together.

If you can't bear all that empty soil around the feet of your roses, and feel as though you're wasting space, and that a little underplanting will keep moisture in the soil and help prevent weeds, rem-ember that reaching through roses to cut flowers or foliage is likely to rip your arm. So, plant your roses in the middle of beds, and perhaps edge the beds with tulips that you intend to leave year-on-year, or with alchemilla, or with herbs with anti-septic qualities, such as mint (which might help fight black spot – see overleaf) or sage, as aphids are reputedly repulsed by the sagey smell.

WHAT WE'VE LEARNED

I will admit that we have put our roses in mixed beds until now, and are in the process of digging them all up and putting them in tighter formation, in keeping with the advice above. It is very difficult to break the habit of making a garden rather than growing for cut flowers. But cutting roses for a wedding is a slow process if, in order to do so, you have to walk endlessly from one end to another of a beautiful mixed border.

Don't just grow 'wedding' roses, or even just long-stemmed roses. This one, 'Wild Edric', has vicious thorns which make it difficult to cut, but five in a wedding scheme can be transforming.

But what about black spot?

Black spot is a fungal disease. It overwinters on fallen leaves and produces new spores to infect new rose leaves each spring. However, the disease can be managed. To prevent black spot:

- Strip infected leaves from both the stems of cut flowers and the plant, and burn them. Never compost black-spotted leaves.
- When the roses have lost all their leaves in winter, rake the leaves up and burn them.
- Give the roses a good mulch of well-rotted manure after you've raked up the fallen leaves and before the next leaves start growing – a good time to do all these jobs is during your midwinter prune. When the next rain falls it won't bounce black spot back up on to the

buds or newly emerging leaves: the mulch will act as a protective layer on top of old spores. When I say 'good mulch', I mean several inches thick – you want to cover the mould spores on the ground effectively as well as feed the roses.

Planting roses

Order your roses bare-rooted in the autumn from a good rose nursery. Even if you want to start with just ten or so, do ask for a discount: nurseries often have multi-buy deals for fewer plants than you'd think. Tell the nursery what you're doing and ask their advice: they will almost certainly be helpful.

Tell them also when you want your roses delivered. As with most things, it's worth ordering in plenty of time, but always think ahead to when the ground will be ready. You can order in October to make sure you get the stock you want, but if you're not going to be ready for them until February, ask for the plants to be delivered then.

Plant your roses in humus-rich beds. Mix a handful of rockdust into the bottom of the hole before

Mycorrhizal fungi

Mycorrhizal fungi are the tiny white fluff on roots, which often look as though they're part of the plant's roots. They form a mutually beneficial relationship with the plant, helping it to access water and nutrients from earth beyond the reach of its own roots. Dipping a shrubby plant or tree's roots into a mycorrhizal inoculant gel before planting can make a real difference to the time it takes to settle in and establish.

planting, and dip the roots of your rose in mycorrhizal fungi. If planting roses bare-rooted in the dormant winter season, then they'll just need heeling in well and a good mulch of well-rotted horse manure to help them settle, in addition to the mycorrhizae. All roses are grafted on to rootstock, and you can see clearly where the graft has been made at the bottom of the stem. Make sure you plant the rose deep enough to cover the graft, which should prevent suckers shooting from the rootstock. If you've made new beds with new soil, you may find that the soil settles around the rose during the first season and the graft node starts to show. Mulch your roses to re-cover the grafted part if this happens.

Do as I say, not as I do. . . I challenge you *not* to pick flowers off newly planted roses for their first season, but to just let them settle in. I have *never* been able to resist cutting roses in their first year, though I know that the plants will be much happier and produce better for me in the second year if I leave them alone.

Pruning

Again, remember that you're not growing roses as part of your garden scheme but to produce lovely long cut stems for floristry. So your pruning schedule might be a little different from when you were just growing for a great show in the garden. There are two key points to remember:

* Roses pruned before Christmas will flower earlier than those pruned in January or February. Gardeners often give their roses a light pruning/tidy-up in early November to prevent wind rock, and then prune properly in February. If you are growing roses for cut flowers and have a collection of, say, ten *Rosa* 'Compassion' (one of my favourites), you could hard-prune half of them in late November / early December, and save the hard pruning of the other half for February, so staggering the main flush of flowers in early summer.
* Staggered pruning in winter will not only give you successional flowering in summer but will also mean that at the end of July you can take your time giving each bush a good cut-back after its first flush, to promote new flowers in the autumn in successional order too. This way you can have roses in flower all summer long.

Grow blowsy roses that your customer can't get elsewhere. This variety is 'Buff Beauty'.

Which roses to grow?

Your first priority is to supply something different from single-stemmed, unscented imports. You can't compete on price where they are concerned. Remember, you're competing with billions of stems exported by the big growers to worldwide destinations every year. You need to grow what won't travel well; what the long-distance trade won't bother with.

You can, however, compete on shape, luxurious-ness and scent. So take instead the market for bobbing heads and thousand-petalled faces: the market which likes a fresh, garden-grown rose. Grow for scent, for blowsiness, for romance.

I'm not going to speak of shrub versus floribunda here, therefore – I'm going to talk about looks. You have so many hundreds of roses to choose from. Your choice will of course depend not only on the market but also on your personal taste, as well as on the space you have available, as this may include the face of your house, the old apple tree in the garden, a bed in the shade of a north-facing wall. . .

Give yourself time and ignore old assumptions

The varieties of rose available within each categ-ory is so wide that I suggest you give yourself time, a budget and a number of catalogues, and go into the field with an open mind. Don't just look at shrub roses, hybrid teas, floribundas. . . Think of your whole plot, look at the ground you have, and think about where roses for cropping will do best. Do you have the space/budget for a whole bed? Or will you only want one or two plants for accent flowers throughout the summer?

Remember that the catalogues will help you concerning whether a rose will flourish on a dry bank, for example, or in damp shade. You are looking for colour, scent and a look which, even if the rose were on its own in a bud vase, is unmistakably 'fresh cut country garden'.

Do look at old rose varieties. Despite their vicious thorns, you might grow a few of them because their scent is intense and they're really hard for florists to find for weddings. Their floppy, velveteen petals give them an air of Edwardian ladies' picture hats; their fluff of central yellow stamens stop a bouquet being too pink or white. One of these roses will go far in a scheme; ten of them are transforming. They also make wonderful hips for use in the winter – if you get to them before the birds, that is.

Consider what the market wants

Think of your market. By all means grow only bright yellow, dark red and virulent pink roses: they'll look stunning in bouquets. But those three colours are not the hot choice for brides, whatever the fashion. Bridal trends tend to undulate within the infinitesimal difference between a sugar-pink and a ballet-shoe-pink rose; between whites and off-whites. Brides sometimes like a little yellow, but more often not. The bridal market is probably half your market – whether you're selling at farmers' markets to do-it-yourself brides, growing for a particular wedding, or growing for your own floristry. Pale-coloured roses will look lovely in *all* the bouquets you make through the year – whereas bright yellow, bright pink and bright red might be a difficult sell to 50 per cent of your market.

Roses are in great demand for wedding flowers. Here we have 'City of York' and 'Madame Alfred Carrière' making romantic flourishes in jam-jar posies for a wedding.

Fresh petal confetti is a great way to make use of the petals of blown roses.

Roses as hedging

Some roses make great hedging plants. One example is the lovely *Rosa mundi*, supposedly named for Henry II's mistress Rosamund Clifford, who is said to have lived in a tower at his hunting lodge at Woodstock, surrounded by a bower of this old pink-and-white-striped rose. (Henry II is supposed to have made a garden at Woodstock, called a 'gaerde'. Inspired by this, one of the orchards here at Common Farm Flowers is called 'The Gaerde'. It is a wild space, with dog roses and irises, hellebores, fruit trees, and wild flowers.) There's a great example of a *Rosa mundi* hedge at Hidcote in Gloucestershire, where it surrounds the nursery beds at the back of the garden. It would make a sturdy windbreak and a good source of short-stemmed cut flowers. And any rose that doesn't mind being clipped into a hedge shape with a hedge cutter isn't going to whine and moan about being pruned.

Climbers and ramblers

The difference between climbers and ramblers? Climbers will repeat-flower; ramblers will flower only once in a season. However, a rambler such as 'Rambling Rector' will gallop happily through an apple tree, flower once with glorious heads of many small, cream, yellow-centred flowers which smell astonishingly of apricots, and then produce enough gorgeous teeny red hips to dress a great many Christmas wreaths. So its single, scent-drowning flowering is not to be sniffed at when its second gift of hips can be so valuable later on.

In a confined space, use your uprights for your roses: send them over pergolas, over gate entrances, through fruit trees – make your space work doubly hard for you. Imagine your cut-flower garden, old and established in 20 years' time, your admirers standing back and thinking how clever you were to not only plant a crab apple 'Red Sentinel' (a prolific flowerer which then makes gorgeous fruit, which sit happily on the branch well into the winter, so is useful for those wreaths again) but to grow through that a 'Rambling Rector' rose. All using a ground space of less than a square metre (10 square feet).

My top five roses to grow for cut flowers

The choice of roses is almost endless. My five recommended ones include 'Compassion' and 'A Shropshire Lad', which are my favourites as much for sentimental reasons as for business. Many of these roses have different registered names (shown here in brackets), though the preferred selling names evoke them best. Growing flowers for cutting is hard work, and you must keep it as pleasurable as you can in order to inspire you to keep at it. So grow the roses whose scent you can conjure on a dour day in February; the roses that remind you of your grandmother, or the house you used to holiday in when you were little.

'A Shropshire Lad' ('Ausled')

Heavy-headed, many-petalled, faded peachy-apricot in colour, this is a difficult rose to use in arrangements, as the sheer weight of its petals pulls the flowers over. It is still one of my top favourites though. It's the very opposite of the rigid, upright, narrow-headed roses that are such a familiar sight in the shops. Floristry is a craft and an art – use your skill to create designs in which such heavy-headed roses feature, and your customers will love you for it.

'Bonica' ('Meidomonac')

As ever, I appeal to you to banish any garden snobbery: when a plant is easily available in garden centres and frequently found in happy gardens, it is because that plant is a reliable performer in many circumstances, including adversity. Bonica can be found for sale everywhere, and will flower against a hot wall or in a cool corner with equal enthusiasm. It's not the most highly scented rose, but what a worker! As an understorey to all your other roses; as a single stem of seven or eight flowers and buds in an arrangement for a wedding or a dinner; just three stems mixed with sweet peas, ox-eye daisies, alchemilla and mint in a bouquet – Bonica is a reliable, quite intense sugary pink, which you will always find use for. And it repeat-flowers constantly from June until October.

'City of York' ('Direktör Benschop')

This repeat-flowering climber is white with pale-lemon buds, and a pretty yellow middle when open. It's prolific for us on an dry-footed, wind-lashed north-facing wall that appears hopeless for plants. It may do better in more favourable conditions, but it is very good for a difficult corner. We cut it all summer long. I love its lemony scent, and the way the buds are pale, pale yellow, developing to a lovely clean white as the flowers open.

'Compassion'

A perfect pink with yellowish centre, this is a generous, repeat-flowering, single-stemmed, highly scented delight. I love a slightly yellow middle, as yellow will lift a pink scheme, which without a lighter touch can be surprisingly flat.

'Munstead Wood' ('Ausbernard')

Choose this for deep-purplish reddish-pinkish colour (you won't only be supplying flowers for weddings) and its rich, fruity scent. This is a romantic rose: a flower of pure velvet.

Here, 'Compassion' with sweet peas and a single lemon-coloured annual chrysanthemum make a very 1950s look.

Cutting and conditioning roses

Flowers you cut for sale or floristry need to last longer than those you cut for the house. A rose in full bloom is lovely in the garden, and may last a fleeting day in the house, but will be too 'blown' to sell, so you need to condition them carefully.

Cut roses when you're beginning to see colour in a fat bud, when the first petals are just beginning to pull themselves open.

Strip the stems of thorns as you cut the roses from the bush. Not only will this save your hands and those of your customers, but also, when pulling one thorny stem at a time out of a bucket of roses for floristry, the thorns will inevitably catch on

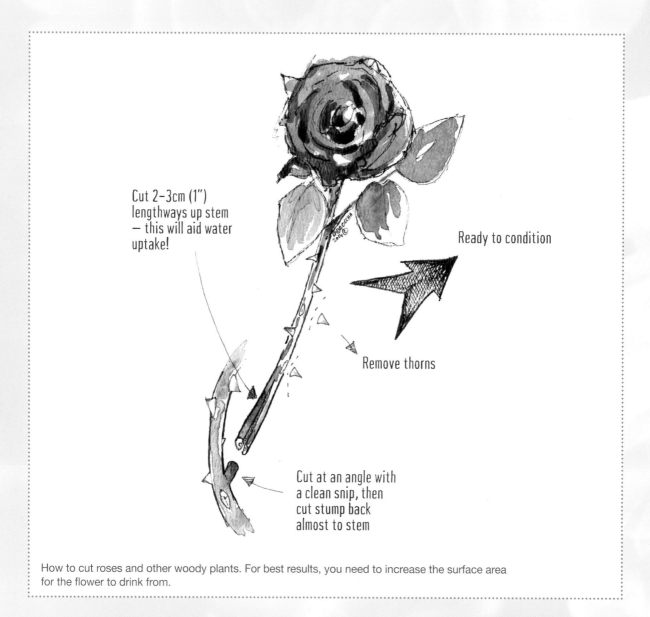

Cut 2–3cm (1") lengthways up stem — this will aid water uptake!

Ready to condition

Remove thorns

Cut at an angle with a clean snip, then cut stump back almost to stem

How to cut roses and other woody plants. For best results, you need to increase the surface area for the flower to drink from.

You should never bash woody stems, as the resulting mess makes it easy for bacteria to breed in the water. Always snip woody stems cleanly.

other thorny stems, and you'll end up with all your roses in a great tangle all over the table. This is annoying enough with single-stemmed roses, but with many-headed roses you'll be cursing if you leave the thorns on.

Remove thorns with scissors. You can buy special yellow plastic bumpy pads to rub thorns off, but I find that they are ineffective and make more mess than simply using scissors for the same job.

Cut roses from the plant in the cool. People ask me how to stop their roses from wilting when they cut them. I don't suffer from wilting roses. I cut them early in the morning or after the sun has gone.

Cut the stem at a steep angle and split woody stems to a length of about 2-3cm (1") from the bottom with a snip of clean, sharp scissors, to allow the rose to drink. Then tidy the bush you've cut from by taking the stump back to just a few millimetres from the stem.

You should never bash woody stems, as the resulting mess makes it easy for bacteria to breed in the water. Always snip woody stems cleanly.

Searing

If your roses do wilt, then you could try searing the wilted stems. Cut your rose stem afresh at a sharp angle across the stem, and snip up the centre of the stem as before. Put it immediately in 5cm (2") of boiling water and leave it there for 30 seconds. Then fill up the container with cold water and leave the rose to recover.

Five top tips for growing roses for cutting

❀ Choose roses to grow for their scent and interesting heads. Remember that the picture people have in their minds when they think of garden roses is one of voluptuous abundance, so grow blooms to satisfy that longing.

❀ If planting roses in new beds, plant more deeply than you might think: in a newly made bed the earth might settle and expose the graft node, from below which suckers might shoot, which you don't want.

❀ To avoid disease, mulch well in winter and burn all affected leaves.

❀ Underplant roses with sage (aphids don't like the smell) and with mint (which has antiseptic properties).

❀ Spray with very weak liquid-soap-and-water solution to fight aphids, in the couple of weeks before the ladybird larvae hatch. Spray every two or three days when the sun is not on the plants (liquid soap + sunshine = scorched leaves).

Chapter eight

Dahlias

There will be those among you who snort at the idea of growing dahlias. Well, I love them, almost as much as I love sweet peas. They are the mainstay of my late-summer cut-flower crop – both workhorses and cancan dancers – without which my business would struggle. Go on, have a look: I bet you'll be surprised at how much dahlias have come on in the past few years. . .

Almost infinite in their variety, dahlias are our next-favourite flowers here at Common Farm, after sweet peas. If sweet peas are like French knickers, dahlias are the cancan girls of the garden.

Tender, tuberous perennials, originally from Mexico and Central America, dahlias were first introduced into Europe in the late eighteenth century. Once associated with a certain kind of obsessive gentleman who grew massive perfect pompons with which to claim victory on the show bench, dahlias have had a resurgence in popularity of late, and deservedly so. While they're always extremely frost-tender, they are otherwise sturdy, prolific flowerers, and with good management can be the mainstay of a cut-flower garden from August until the first frosts.

How to grow dahlias for cut flowers

Most of my planting here at Common Farm is driven by the vagaries of each season's weather. In contrast, I have strictly timetabled 'dahlia' days, which are diarized years in advance and around which my dahlia growing is built. We lift the tubers after the first frosts (mid-November), store them until 1 April, when we get them out and pot up the tubers, and we plant out grown-on plants on 1 June. It's important to be matter-of-fact with dahlias, especially if you're dealing with a lot of them.

Ordering cuttings

In October, have a browse through the websites of a couple of good dahlia cuttings suppliers. Give yourself time: there are almost as many dahlias to choose from as there are grains of sand on a beach. See pages 133-5 for a summary of the different kinds available.

Dahlia plantlets just arrived in April from the supplier. Pot them up quickly, so that their roots will rehydrate and they'll settle in well for planting out in June.

Make a wish list, then edit that list. Thirty dahlia cuttings will give you a huge dahlia patch for a very reasonable capital outlay and, if the plants are well looked after, you'll get literally tonnes of flowers from those plants through the season. Order before Christmas: dahlias are extremely popular, and people order new every year to satisfy their desire to change the colour schemes in their gardens. Suppliers do run out of popular colours and shapes, so get in early to be sure of having the scheme you've chosen.

The growers send out rooted cuttings in early April, so if you order in September you'll have a long time to forget all about them. It will be an occasion of great excitement when those boxes of plantlets are delivered to your door.

When your dahlias do arrive, hurry to get them potted up. The cuttings will very likely arrive in sweaty plastic boxes, which are ideal for travelling in, but the baby plants won't survive in them for very long.

Mix your compost with a little grit: you want it to be free-draining, so that the new roots won't rot off. Then put the pots of plants under cover. It's only early April, and dahlias are extremely susceptible to frost. Plant them out with all the rest of your tender plants, when all risk of frost is past.

Propagating dahlias

Dahlias are relatively easy to propagate. If you have the space and warmth, get the tubers out from their winter storage (see page 132) in February/March, and, having cut out any withered or mouldy parts from last year's root system, lay them on a layer of compost on a heated propagating bench (old mushroom compost trays make perfect containers – one dahlia per tray). Then lightly cover (not as deeply as if you were planting) the tubers with compost, making sure it is free-draining. The tubers will sprout away happily, so long as they're not too wet.

The greenhouse or polytunnel must be frost-free. If you have a cold greenhouse or are unable to heat your propagating bench, then don't try propagating dahlias under cover until April – and note I still said 'under cover'.

In three or four weeks, your tubers will reward you with little shoots around the original tuber central stem. Shoots of 7m (3") long can be sliced away from the original, last summer's stem, as basal cuttings to root on in 22cm (9") pots – still on that warm bench if you have room. You can dip your cuttings into a rooting powder if you like. Be careful not to over-water – you wouldn't want carefully taken cuttings damping off!

How we propagate dahlias

I will admit that at Common Farm we propagate dahlias in a more unconventional, but no less successful manner. We don't have many spacious warm beds for propagating early shoots, so we don't get our dahlia tubers out of storage until early April.

We examine them carefully for rot and for shrivelled, dried-up tubers, and cut those parts out. We then cut up the tubers into sections with a sharp kitchen knife, where possible leaving out last summer's dried-out stem. By this time of year, you can see the tiny, moonstone-like eyes of the dahlia shoots beginning to form, and so it's easy to cut up your old tubers to make several new plants out of each.

We pot our cut-up dahlia tuber sections into a 50:50 mix of municipal green-waste compost and garden grit, give the pots a good water, and line them up in the polytunnel – which, although unheated, by April is warm enough to protect from frost. Our plants shoot in two or three weeks and we pot the plants on once or maybe twice if they grow quickly.

When to plant out

As a cut-flower grower, you won't need your dahlias to be flowering until mid- to late July, because you

A note of caution

If following my technique for propagating dahlias, *be careful*! Knives are sharp; tubers are sometimes tough to carve up. Use your common sense and put the tuber down on a hard surface and cut it as if you were cutting up a hard-skinned squash – with your fingers kept well away from the blade and preferably with gloves on.

Dahlias, potted up in half grit, half compost, with a handful of well-rotted manure for feed, ready for planting out on 1 June (not before!).

will have carefully planted your cut-flower patch to be successfully flowering throughout the season. So don't risk your dahlias for the sake of a few early blooms. Plant your dahlias out into beds on or around 1 June – or when you are absolutely sure there will be no more frost.

Disbudding

You *can* 'disbud' selected side shoots on budding dahlia flowers (as shown opposite) in order to encourage super-blooms. This is great if you're looking to produce prize-winning flowers for comp-etition, or if your buyer is looking for very big flower heads on very strong stems. If, however, you're growing for flowers to be turned into bou-quets, you may think twice about disbudding in this way. Dahlia stems can be very thick, making them difficult to include in hand-tie bouquets. If you leave the side shoots to grow into flowers, you'll have slimmer stems and more, slightly smaller, flower heads and it'll be easier to make bouquets with them. The way you grow your flowers is a great deal about how you're presenting

them to the market: if you're growing for a local florist or a wholesaler, ask which they prefer; if you're growing for bouquets, try both systems and see which you prefer.

Whether or not you choose to disbud your dahlias, pinch out the first, leading flower spikes as they develop, to encourage strong side shoots and a lovely bushy plant producing masses and masses of flowers.

Staking dahlias

Stake your dahlias as soon as you've planted them out: if you wait, you'll be making a much more difficult job for yourself later.

Depending on the number you've planted, you can choose one of several ways of supporting them. Some dahlias have very thick stems, but they grow so fast in the season that those stems tend to be brittle, and in a high wind you can lose whole plants, snapped off at the root in a storm. Your dahlias will keep flowering for you until the first frosts, during which time you may have had vicious, thunderous squalls in August, and the first of the autumn storms in September and October, so staking is essential if you're to be rewarded for your efforts with a good dahlia crop.

Each dahlia planted individually in a mixed scheme can be given a teepee of stakes to support it. Tie the teepee at the top and make sure the stakes are driven good and hard into the ground. The dahlia will grow through the stakes and be glad of the support on a windy day.

A whole bed of dahlias can be supported by a horizontal layer of pea netting, laid across the bed at a height of about 1m (3') and tied to stakes along the edges, as described in Chapter 1 (see page 23) and illustrated opposite.

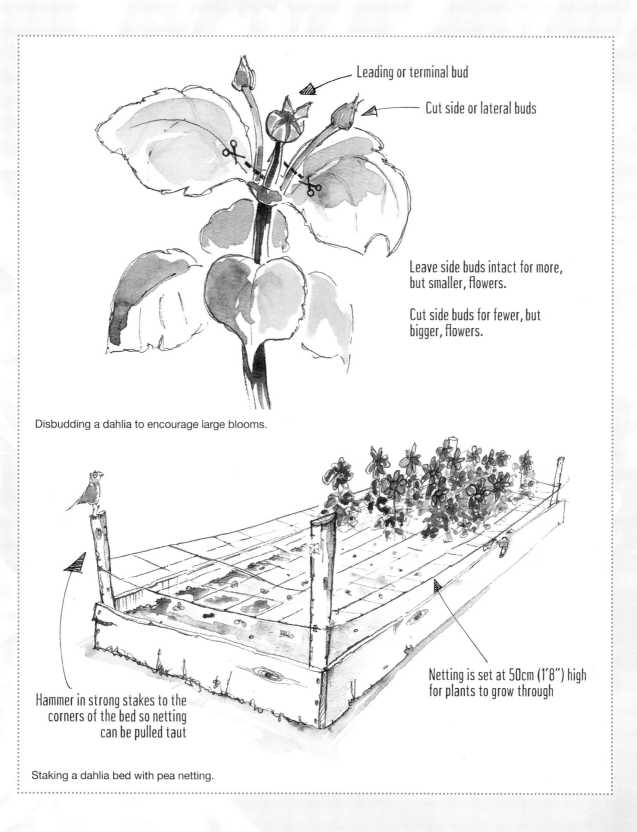

Leading or terminal bud

Cut side or lateral buds

Leave side buds intact for more, but smaller, flowers.

Cut side buds for fewer, but bigger, flowers.

Disbudding a dahlia to encourage large blooms.

Netting is set at 50cm (1'8") high for plants to grow through

Hammer in strong stakes to the corners of the bed so netting can be pulled taut

Staking a dahlia bed with pea netting.

If you forget to install this netting system in time, and your dahlias are suddenly too tall to net without bruising, snapping and generally causing more damage than not staking would cause, don't panic. Put stakes at the four corners of your dahlia bed (and a few times in between if the bed is long), then make a zigzag of string or twine between the stakes, snaking the string between the shooting dahlias, again at about 1m high. The greater a 'cat's cradle' effect you make, the more effective the support. When you've made your cat's cradle, be sure to take your string all the way round the outside of the beds, encircling the stakes and preventing the plants from falling outwards in a high wind as well as from collapsing upon one another. Make sure the zigzag of string is pulled tight and that the stakes are jammed firmly in the ground (this is about support, not knitting).

Feeding

Dahlias that are expected to produce fresh flowers for picking daily will need a great deal of feeding.

Compost tea will feed the ground so that they have nutritious soil to grow in, while a good nitrogen feed such as nettle tea will promote strong growth and help your dahlias survive slug or aphid attack.

A potassium-rich feed such as comfrey tea, applied weekly throughout the flowering season after mid-June, will promote the formation of lots of great big (cash-flow-generating) flowers. See Chapter 1, page 28, for recipes for these liquid feeds.

Lifting and storing dahlia tubers

You can leave your dahlia tubers in the ground to overwinter, if your drainage is good and the frost not too hard. Damp is as much the enemy of the dormant dahlia as is cold. If you plan to leave

them in the ground, then do plant them with a good scoop of grit mixed into the soil, so that they don't end up sitting in a bog. Equally, if overwintering in the ground, they'll need a good mulch of well-rotted manure or garden compost several inches thick to protect them from the frost. Dahlias left to overwinter outside will send out new shoots in the spring in the same way that dahlias potted up on a warm bench will. You can take cuttings of the new shoots, as described earlier (see page 129) at that time.

If you're lifting your dahlias to store out of frost and damp, wait until the first frost bites, then arrange them on to a flat, dry, frost-free space and leave them to dry off for a few days. Take off the majority of the soil stuck to them if you can, but not quite all of it. Then they're ready for storing in a place where they'll be kept dark and frost-free, and where damp won't encourage rot.

We store our dahlia tubers in crates in a dry barn, heaped on a bed of polystyrene blocks and covered with a very tattered, old 100-per-cent wool (and

We store our dahlias in our barn, in crates, on polystyrene blocks, protected by an old 100-per-cent wool rug.

therefore totally breathable) rug. From time to time, on a warm day in winter, I take the rug off and let the tubers breathe. We don't live in a cold area, though: if you do suffer really cold winters, then an old duvet makes a good cover, or straw – but watch out for nesting mice.

Which dahlias to grow?

Think in advance what dahlias are going to do for you. Are you going to sell bunches of five stems at a farmers' market? Will you have them in bunches at the garden gate? Are you growing for wedding flowers? Will your local florist buy from you? Very heavy heads might be easily bruised and squashed in transit. Very flat-faced dahlias might be too fragile to keep their petals during their journey from plot to customer's table. I can tell you how and why I make my choices: use my decision-making

process to help you devise your own, and then choose your dahlias accordingly.

I think the way to grow dahlias is to be prepared to discard, give away or sell on a great many plants, and often, otherwise the flowers you offer year-on-year will be samey – while your customers' tastes will be driven by the colours that are fashionable that year. It's worth bearing this in mind when making your business plan. Fashions change and, especially if you're growing for wedding flowers, it's a good idea to be up with the trends (see box on page 135). So, you will change your dahlias from time to time: budget for it.

When ordering, think carefully of the colour combinations you're going to be selling during the second half of the season. Don't be too matchy-matchy, but remember you're going to be putting these dahlias together with the rest of the contents of your cut-flower patch to sell, so they need to work together in order to attract your bee-like customers. Think of the colours your sunflowers, roses and hydrangeas are, for example, and choose your dahlias accordingly.

Types of dahlias

Perhaps this list of different dahlia shapes will begin to help you make your choice. Perhaps. . . Though the only ones I wouldn't choose for cut-flower production are the dinner plates and the collarettes – *all* the other dahlias (apart from dwarf varieties) should be fair game for cutting I think.

Pompon, ball and miniature ball

With almost-perfectly spherical heads made of

I think the way to grow dahlias is to be prepared to discard, give away or sell on a great many plants – and often.

hundreds of tightly packed little half-furled petals, pompon dahlias are perhaps the type most associated with those obsessive show-bench-competing gentlemen I mentioned at the start of this chapter.

The huge pompon heads might be difficult to use in everyday floristry, but there are lots of smaller examples which work really well in bouquets, and the tiniest of them, wired, make wonderful buttonholes or flowers for hair.

Don't assume, when looking at a catalogue, that a pompon dahlia is a large, cumbersome head. Ball dahlias are the really big round heads, and there are also miniature balls, which make good cut flowers. Not all suppliers categorize their dahlia varieties the same way, so if you're unclear about how big the heads are on the dahlias you're thinking of ordering, do ring the supplier: they'll tell you how big the heads get, or whether they have a similar, miniature variety if you'd like one.

Dahlias aren't expensive, given the number of flowers you'll get off a plant in a season, so I recommend you have a play with one or two ball and pompon varieties: you may find them more useful, interesting and hard-working in your cut-flower patch than you might assume.

Collarette

These have flat, daisy-shaped flower heads with only eight or nine petals. If you want to grow wildlife-friendly dahlias in your garden, these are the best for foraging bees. The pollen is easily available – presented on a plate, as it were, by the flower, which frames the pollen with bright and attractive (to the bees as well) petals.

WHAT WE'VE LEARNED

Collarette dahlias are *not* the best flowers for cutting, as the petals can be easily bruised, and also tend to fall off within a few days of cutting.

Cactus

With flower heads made of hundreds of spiky-shaped petals, these make dramatic cut flowers. Though the pollen hearts are hidden deep inside the flower head, the bees will find them, so when

A cactus dahlia just beginning to open. This is the perfect time to cut it if it's to last five days or longer in the vase without chemical preservatives.

cutting the flowers early in the morning watch out for soporific bumble bees who've overnighted in a dahlia hotel and may be taken by surprise by their berth suddenly moving.

Semi-cactus

A softer version of the cactus, the petals of the semi-cactus aren't so spiked, and you may think they make a better cut flower: less sharp-edged; perhaps easier to sell. We find that the semi-cactus lasts better in water than the cactus, but we grow both for sale as cut flowers.

Dinner plate

These vast flowers are ideal contenders for the competition bench. We have one or two plants just for fun, but they don't make great cut flowers, as the heads are so huge they're almost impossible to arrange.

Fully out, these dahlias really do make the size of dinner plates, or even a medium-sized wedding hat. Unless you become a serious dahlia grower supplying markets where florists who are dressing huge events are buying, then you might hold back from the dinner-plate varieties. You'll get a great deal more from manageable-sized flowers that are much more versatile.

Waterlily

These beautiful dahlias truly are waterlily-shaped. Many of the Karma series, which have a reputation for being the best-lasting of the dahlias as cut flowers, are waterlily-shaped.

Decorative

Softer than a pompon/ball, but still with hundreds of petals making a fat cushion of a flower, the decorative is on its way to being a cactus or a waterlily, but is not quite either. It's usually a good size for cutting – not too big, and therefore good as an accent flower in floristry.

A note on fashion

The kind of person who likes the idea of flower farming for a living is possibly the kind of person to be horrified at the thought of being influenced by anything so ridiculous as fashion. But the fact is that your customers *are* influenced by fashion, and will make their demands accordingly.

Talk to any branding or marketing guru and he or she will shock you with a list as long as your arm of how we are all influenced by fashion. And we cut-flower growers, dealing in a product of such fleeting beauty, have to allow ourselves to keep up.

One of the reasons the cut-flower industry has become so moribund is that farmers have been growing what is easy, cheap and disease-resistant, rather than what is desired and goes with the houses and colour schemes in which the flowers will end up.

To your customer, the cut flowers you're growing are expensive luxuries: gifts with which they will make special treats. If the fashion designers in Paris spend time looking at next year's pantones before they cut their cloth, then who are we gardeners to sniff at fashion?

Don't be put off: a moribund industry is one to get into if you're prepared to look at *why* the industry's become so hidebound, and if you're prepared to kick over the traces a bit!

Dwarf

Although good for bedding schemes, dwarf dahlias don't make brilliant cut flowers because of their short stems. Whether selling per stem or growing for your own floristry, you'll need longer stems than those offered by the dwarf varieties.

My top five dahlias to grow for cut flowers

The following varieties are not the longest-lasting in the vase – but they are my best dancers in a bouquet. I grow for bouquets and weddings, so that's what I'm thinking of when I plan my scheme. Again, you may have very different criteria, and it's worth having a clear focus before you start making your choices.

'Apricot Desire'

I'm a big fan of apricot, and a touch of lemon at throat and wrist really tickles my floristry buds, so this is a must-grow for me. It's very good mixed with whites and silvers in a wedding scheme.

'Barbarry Olympic'

This lilac ball dahlia is irresistible to me. A good ball dahlia feels like a really good joke. They never fail to make me smile; I sometimes even laugh out loud with delight to see them.

'Blackberry Ripple'

A gorgeous striped purple-and-white decorative dahlia, this is the strongest of our collection and easiest to propagate. It sometimes reverts to producing pure purple or pure white flowers. It's a very good purple when it does that, though my mother disagrees and says they're 'putrid' – you see what I mean about tastes being different!

'Orfeo'

This rich pink, big cactus dahlia has long, strong stems and lasts a good five days when cut.

'Summer Night' ('Nuit d'ete')

A gorgeous dark, dark blood-red cactus dahlia. I love all the deep-dark dahlias. They work really well as the season wanes, revealing a dense palette of deep, velvety colours within each showy flower.

Cutting and conditioning dahlias

Dahlias aren't the longest-lasting of cut flowers. It's worth explaining this to your customers when they buy their dahlias, so that they will enjoy them for their fleeting glory and won't be disappointed.

Dahlias cut when they're just opening will amuse. Here's 'Apricot Desire' pinging into flower, one petal at a time.

Cut the flowers early in the morning, direct into clean water, and their hollow stems won't have time to dry out and slow down the capillary effect of their drinking cells. They will give you five to seven days in water, so long as the water (and vase) is kept scrupulously clean. Dahlias will make water dirty, and their stems can go brown in water. In hot weather, a tiny drop of thin bleach or white vinegar in the water will help stop the development of flower-killing bacteria.

You'll see that my advice here differs slightly from that of Richard Ramsey (see page 139): he is growing for sale at market, whereas my flowers reach my end user more quickly. Do try both methods. (You'll almost never be given exactly the same advice from two horticulturalists. But you will be running *your* business from *your* garden, so experiment and see what works best for *you*.)

Some growers cut their dahlias quite open and treat them with a silver-nitrate solution to make them last longer. We don't treat our dahlias with any flower-preserving liquids. Instead, we cut them when they're perhaps only a third open – they open quite happily in the vase – and we sell them when the flowers aren't fully out, so that our customers can enjoy their changing shapes day by day until they finish a week later.

Buckets of dahlias conditioning in a barn may be populated by lots of sleepy bees – and earwigs. I recommend a good shake to rid your flowers and bouquets of too much wildlife: your customers may love your green credentials, but that doesn't mean they want earwigs crawling about on their kitchen tables, or bees suddenly flying at them as they arrange their bouquet in their vase room (vase room! Imagine a vase room. . .) Also, bumble bees prefer to stay relatively near their nests – taking them to market or sending them far from home is just unkind.

Bouquet combinations with other flowers

Dahlias are great bouquet flowers, and worth a lot if you're growing them to sell as individual stems to florists/wholesalers or in your own floristry. Their velvety petals have a lovely texture, which only enhances their gorgeous array of colours.

When they first start to flower they can seem to clash with the more delicate early-summer blooms that are still flowering away. Their colours seem harsh, sometimes gloomy, when they begin to flower in July, but they intensify as the season goes on and the light changes. Even the dark dahlias almost seem to glow on a grey, dank, cloudy October day. In the glaring high sun of July and early August, it's possible to look at a particular dark dahlia and wonder why you grew it: its colours seem flat, and against your still-flowering roses and sweet peas it just doesn't compete. But by late September or early October, when the sun is lower and the grey clouds make a deeper background, even the darkest dahlia is transformed by revealing a second storey of brighter reds and vermillions; a heart as crimson as Chanel's Rouge Noir nail varnish.

Make no assumptions that dahlias will be obedient to your will. They will work with sunflowers, roses, etc., but if you're making posies or bouquets for sale with them, don't assume that they will do as they're told. Their heads are top heavy, and their stems can be thick: they're fiddly to manage in floristry.

They love to be framed with umbellifers such as ammi, wild carrot and fennel (the mustard colour of bronze fennel is fantastic with a lot of dahlia colours), while the green spikes of bells of Ireland work beautifully through a mix of dahlias. Verbena bonariensis, so long as you cut it early enough so

A bouquet of dahlias mixed with cosmos, scabious, physocarpus, verbena bonariensis and everlasting flowers.

that it doesn't shed, makes a wonderful cut flower, and its intense blue-purple flower heads make lovely little cushions in a bunch of dahlias.

We use echinacea when it's not out yet as a button flower in a dahlia mix, and often, when the echinacea is fully flowering, we tear the petals off and use it as a thistle shape in dahlia bouquets – the texture mix is striking and very successful.

The globe thistles and sea hollies are great with dahlias, and our pale-blue perennial aster makes a lovely sprinkle of stars across a dark dahlia background. Sedum can be large and heavy in a dahlia mix – it depends when you last split your sedums whether you get small enough heads to be useful in everyday bouquets (see Chapter 4, page 80).

Five top tips for growing dahlias for cutting

❉ Be prepared to ring the changes often: fashions change, and your customers will be happy if your dahlias change with them.

❉ Never plant a dahlia out until after the last frosts.

❉ Pinch out the leading flower-bud shoots to create bushy, floriferous plants.

❉ Stake dahlias early in the season to avert damage from rain and storms later on.

❉ Lift dahlias after the first frosts, unless your local climate is not too wet or cold, and store them in a dry, frost-free environment. Check them over the winter for signs of mould and rot, and if you spot any, cut it out to prevent it spreading.

A few words from Richard Ramsey of Withypitts Dahlias

Richard Ramsey is a specialist dahlia grower, and a great example of a grower taking advantage of new ways to sell his product. He not only sells his own crop at market and through direct sales, but also sells rooted cuttings to a growing customer base of cut-flower growers.

"Withypitts Dahlias evolved from an increasingly successful hobby activity. Our success in growing exhibition-quality blooms for the cut-flower trade is no accident, as I grew up on a specialist dahlia nursery, learning from the best in the country at that time.

Dahlias are perfect cut flowers, providing a wealth of colour – from vibrant jewel tones to subtle pastel shades, but not blue or green. From tiny pompons to giant 'dinner plates' of over 25cm (10"), and a variety of shapes from balls to spiky 'cacti', the textures enable their use in any setting or occasion.

Generally I would always recommend the small decorative waterlily type, for example 'Taratahi Ruby', 'Carolina Wagemans' and 'Glorie Van Heemstede'. Disbudded, these will reach a bloom size of 15cm (6"); un-disbudded, around 10cm (4"). They are prolific, produce blooms on long, strong stems and have excellent lasting qualities. Pompon types are simply fantastic for table decorations or buttonholes, and the larger-flowered types, like the big cactus dahlias, are just amazing in any setting.

Dahlias do not open well if cut very young. Harvest early in the day and place in fresh water immediately. Cut the stems under warm to hot water at 45 degrees with a sharp knife, as scissors crush the capillaries (sorry, florists!). This prevents air seeping into the capillaries and eases the take-up of water. When using in a vase, change the water very regularly and trim a little from the stem as described above. I would then expect blooms to last 7-10 days in a cool environment. I do not find that the Karma range last longer than any others we grow. However, I learn from our designer clients that imported dahlias drop petals very quickly. Another good reason to buy locally grown flowers!"

Chapter nine
Sweet peas

When the original idea for this book germinated a few years ago, it was to be an ode to the sweet pea. These were the first cut flowers I grew in any quantity; the first cut flowers I sold in little bunches in a barrow outside my front door. Sweet peas are the backbone of my cut-flower business – and while we now grow hundreds of different varieties of flower and foliage, it is the first sweet pea of the year that still gives me most pleasure.

'Mollie Rilstone' – competing hard to be my favourite sweet pea variety.

There will be some flower that above all inspires you to want to grow cut flowers in abundance. There will be one whose scent, whose shape, whose texture, whose exquisite perfection is the one you would grow if you could choose only one. For us, this flower is the sweet pea. Indeed, this chapter might be said to be a love letter to *Lathyrus odoratus*. . .

On my desk now I have one called 'Mollie Rilstone'. She's a Spencer variety (see page 149). Her petals have the texture of fine crêpe de Chine, she's creamy white with a ballet-shoe-pink picotee border on ruffles that only the very greatest 1920s lady's maid would have been able to iron into her mistress's Lanvin undergarments.

Good-quality sweet peas are relatively difficult to get hold of and so earn a premium. If you want to avoid competing with cheap imports, then they are a good choice to grow. They have a relatively short shelf-life: they certainly won't survive out of water in cold storage for a week as, say, a tulip might, and so buyers are looking for locally grown flowers which spend less time in transit between field and vase.

Florists often complain that they can't get good-quality scented sweet peas that will last. Quality sweet peas, grown straight, tied in bunches of 25, cut when the first flower is open but the others not, and delivered the same morning to your buyer will beat other imports hands down.

Customers pay up to £1 a stem for long, strong-stemmed sweet peas for a wedding. Bear that in mind when you're offered 25p a stem. . . Yes, that might be from a wholesaler, but if you're selling direct to the florist you should think about what the wholesaler would charge, and whether their sweet peas are as fresh and scented as yours, as untreated with chemicals, and you should price your flowers accordingly. You may be competing against flowers flown in from Holland, already several days old.

How to grow sweet peas for cut flowers

Sweet peas are hardy annuals: easy to grow if you can keep the mice away from the germinating seed. Mice like sweet pea seed more than almost anything else, and they like it best when it is just absorbing water and preparing to burst into life.

Sowing seed

According to conventional wisdom, sweet pea seed should be treated before planting, to aid germination. This can be done in different ways:

WHAT WE'VE LEARNED

We don't pre-treat our sweet pea seed at all. So long as the seed is fresh, it should germinate well without any fiddly extra attentions. We did used to soak it, then one year, in a hurry, I just sowed a tray. And it all germinated. So now, straight from the packet, we sow sweet pea seed dry, uncut, not sandpapered, a knuckle's depth in good, moist, well-drained seed compost in a tall pot, and cover it with a transparent lid until the seedlings have sprouted and grown at least one leaf. At that stage you're relatively safe from the mouse, as once the plant begins to grow, the seed withers and eventually disappears into the new plant.

- Soak the seed overnight to soften the armour-plated carapace of the seed before planting. This will encourage germination perhaps two or three days earlier than with untreated seed.
- Cut the seed along the joining line with a paring knife: this too will weaken the armour plating of the seed shell and encourage swifter germination.
- Sandpaper the seed to weaken the carapace – hard work on something so small!

The quicker the germination, the better – as the faster a shoot and root uncurls from within the seed, the less time there is for the mouse to get it. However, life is short. If you want your sweet pea seeds to sprout quickly, then soaking them is probably the least labour-intensive of all these treatments.

Sow sweet pea seed into deep seed trays or tall pots to give space for the plants' roots to develop. Sweet peas, like all leguminous plants, have long roots, which search deep in the soil for water and nutrition.

If you're planning to sow a lot of sweet pea seed, you might like to try the 'Rootrainer', which is a deep seed tray made from folding plastic modules. Not only does this design provide room for long roots but it also maximizes space in your greenhouse or polytunnel by allowing the seedlings to be grown in a tight space while separated by the modules. The seedlings develop strong root systems, untangled in their neighbours', and so are easy to plant out.

People also recommend sowing sweet pea seed into toilet-roll cardboard tubes, but we find that these tend to get mouldy, and we don't like mould in our greenhouse anywhere!

If sowing directly into the soil outside, sow two seeds for every one you expect to germinate, as at least half your seed will be sacrificed to the marauding mouse. We never direct-sow sweet peas.

Sow sweet pea seed into deep seed trays or tall pots to give space for the plants' roots to develop.

When to sow

Since sweet peas are hardy annuals, they can be sown into seed trays in September to overwinter somewhere cool for planting out in March or April, or for planting into a polytunnel as soon as it has been cleared of the previous season's crops and the seedlings are big enough. Tunnel-planted autumn-sown sweet peas will make a good early crop. If you grow the Solstice varieties, which are bred for early flowering, and perhaps consider a little heat and light in your tunnel, you can start cropping sweet peas as early as February. Our tunnel is neither artificially lit nor heated, and our Solstice varieties, sown in trays in September and put into the ground in the tunnel as good-

sized plantlets in December, will begin to flower for us in April.

You can grow successional crops of sweet peas by sowing three times during the year:

* Sow Solstice varieties for early cropping in September, planting them 'out' into beds in your tunnel in January, or earlier if your tunnel has heat, for cropping from late March or early April.
* Sow a crop in September for planting 'out' into your tunnel in March, for flowers from mid-May onwards.
* Sow a crop in January/February for planting out into the field in April, for cropping from July to September.

October brides often say to me, "I suppose I'll be too late for sweet peas," which is why we plant our third crop of sweet peas in April. I love to see a bride happy.

Clear your sweet pea crops ruthlessly after six weeks in full flower (probably two months since the first flower). It seems counter-intuitive to pull out apparently perfectly healthy cropping sweet pea plants. But if you're looking for good, strong production on long stems, then the plants will do better for you when they're younger. Sow three crops and you'll have one flowering for May and June, one for July and August, and one for September and October until the frosts bite hard.

In a good summer sweet peas will be perfectly happy outside, but if you have space, you might consider growing half your crop under cover.

Pinching out and growing on seedlings

Pinch out the top of all sweet pea seedlings, no matter which variety or when sown, when the shoot has two sets of true leaves, to just above the second set of true leaves. Your seedling will now grow two further shoots from either side of the original, so that it then has three shoots. This will give you very strong, freely flowering plants.

Once the new side shoots have grown, you can choose the strongest and pinch out the other two shoots, leaving only one shoot with leaves on your original seedling, and plant it to grow up a cordon in a polytunnel or greenhouse or even outside in a sheltered area. When the plant reaches the top of the cordon you can untie it, lay it along the ground and start growing the plant up a new stake or string to get more out of it.

Cordon-grown sweet peas from which the tendrils are also pinched out as they grow up produce the longest stems and the biggest flowers, worth the most money per stem. However, this is a very labour-intensive way of growing sweet peas. We neither pinch out tendrils nor even two of the three shoots from each seedling, and we get really good-length sweet pea stems all summer long.

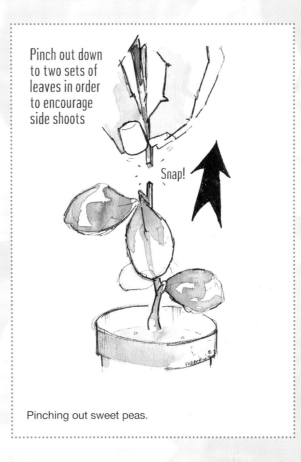

Pinch out down to two sets of leaves in order to encourage side shoots

Snap!

Pinching out sweet peas.

Sweet pea seedlings ready to have their top leaves pinched out, leaving two sets of true leaves to grow on.

Planting and management

Sweet peas like humus-rich soil. They're greedy, thirsty plants, which will suffer from mildew if the roots get too dry. They like their leaves, stems and flowers to be warm and in full sun. This can be a difficult combination to conjure without proper advance preparation of the soil, so it is well worth digging in some rich organic matter: well-rotted horse manure or good, rich home-made compost. Your sweet peas will do much better this way than if just planted into ordinary garden soil. If you're going to grow your plants in a straight line (see overleaf), dig a trench 30cm (12") wide and deep, fill it with organic matter, and plant your sweet peas 30cm apart along it.

October brides often say to me, "I suppose I'll be too late for sweet peas," which is why we plant our third crop of sweet peas in April. I love to see a bride happy.

Sweet peas are massively floriferous: it's very important to cut *all* their flowers *constantly* to keep them flowering for a good long time.

It helps to feed them while they're growing too: compost tea will feed the earth, nettle tea will encourage strong stem and plant growth, and comfrey tea will encourage flowering. (See Chapter 1, page 28, for recipes for these feeds.) An occasional feed with very weak seaweed solution will ring the changes and make sure your sweet peas are given trace elements which might be missing in other feeds. Well-fed plants are much better at fighting off pests and disease, even resisting slug damage, especially when they're planted into well-fed earth. You can also use seaweed solution

as a foliar feed on sweet peas, but be conscious when you use it that it will stain flowers and leaves for a day or two. We foliar-feed on a Monday, which is the day we deadhead the entire garden so that we have lovely fresh flowers as demand grows throughout the week.

Plant sacrificial achillea near sweet peas, because greenfly seem to prefer it. Sweet peas can suffer terribly from those green pests – though if you garden without pesticides, then the ladybirds will keep up the war on greenfly. . . I promise!

Sweet peas also suffer from botrytis during prolonged wet weather. You'll find tiny brown spots appearing on the petals of your flowers soon after they're cut, making the flowers unsaleable. Sweet peas grown under cover suffer much less from botrytis in a wet summer, but you can rig up protection over outdoor crops if necessary.

Supporting sweet peas

Sweet peas are climbers. The skill for you is to grow these climbers up a frame which not only supports them but also makes it easy for you to cut each and every flower as it blooms. Leaving sweet peas to go to seed because you haven't seen, or couldn't reach, a flower head through a teepee arrangement will potentially hurt your cash flow. So think hard about how you'll plant and support them so that you've got maximum access for management of the plants.

For commercial quantities of sweet peas you should think about growing them up a horizontal

Dig a trench and fill it with well-rotted compost and/or manure before planting your sweet peas in it.

Sweet peas grown up chicken wire and underplanted with cornflowers.

support, such as pea netting or chicken wire, in a straight line. The flowers are easier to cut this way than if they're grown up teepee-shaped arrangements, and, perhaps more importantly, developing seed pods can also be found and removed more easily than if they're lost in a jungle of leaves and tendrils inside a teepee structure. However, your choice of support structure will depend on how many plants you are growing, among other factors.

Option 1: chicken wire. This is a greener option than pea netting (see right) because the leftover plant material can be burned off with a flame-thrower at the end of the season and the whole support rolled up and put in the shed, then used again next year.

I will admit this is a fiddly, hand-slicing way to farm sweet peas, but it's practical and 'eco' in a small plot.

Option 2: pea netting. This is the easiest-to-handle support available, lightest for the sweet peas to grow up, and works best. At the end of the season there's no way you can roll it up to use again – well, if you can you're better at handling the resulting tangled mess than I – and unless you wash it effectively in disinfectant you won't be able to remove mildew spores. But it is plastic, and if, like so much plastic, it ends up floating in the ocean, it presents a serious threat to wildlife. If you choose pea netting to support your flowers, think hard about how you will dispose of it at the end of the season.

Option 3: a teepee of canes. You can run string in a zigzag along a long line of canes leaning in to one another, which is efficient, in that you can create your cat's cradle of string so that each plant has its own climbing frame, but a time-consuming way to build a framework if you're growing hundreds of sweet peas. Also, inevitably there will be flowers you lose inside any kind of teepee framework, which will make seed, and so encourage your plants to go over earlier.

Option 4: hazel prunings. If you have a small plot and are growing maybe one row of 20 sweet pea plants, then hazel prunings make great staking material. Hazel grows with one side of the branch flat while the twigs all crowd out on the other,

Hazel prunings make good supports for sweet peas.

which means you can create a very tidy arrangement with it. Of course, as in a teepee, again you risk a tangle of material in the middle of the arrangement in which flowers might get lost, go over and encourage the plant to go to seed.

If you're growing three separate crops of sweet peas to take you through the season, then plants going to seed as a result of some flowers being overlooked might not matter so much. As ever, look at your space and think hard about where you can maximize profits. Do you need to guarantee sweet peas for a bride in October? Or are you planning to sell posies at the gate where your customers will largely get what they're given? In the second case you won't need to keep your sweet peas in such perfect condition all summer long.

Saving sweet pea seed

Sweet peas should come back true from seed unless the variety is an F1 hybrid.

If you're planning to save seed, you might plant a separate stand of sweet peas for seed development. This way you can leave the flowers to go over and the seed to develop for a nice long time during the summer, ensuring that you have good-quality, strong, viable seed for planting in the autumn.

If you plan to take seed from stands of sweet peas that you're also cutting, you risk not giving your seed pods enough time to mature at the end of the season, or shortening your cutting season because you're leaving seed to mature on the plant – which seems to encourage the remaining flowers to grow less vigorously and on shorter stems (the plant thinks it's done its job and no longer makes the effort with its flowers).

Which sweet peas to grow?

Choose carefully: are you growing for weddings? For market? A barrow outside your gate? What will your market like best: scent – or impact?

Types of sweet pea

As is the case in many aspects of life, your dear horticultural friends are inconsistent with their terms when they write about the different classifications of sweet pea. I am a great one for simplicity, and so here I've chopped them into three categories which suit me. Do consult sweet-pea expert Roger Parsons for more detail (see Resources section).

Heritage varieties

Sweet pea flowers were first seen in the UK at the end of the seventeenth century. A plant probably closely resembling that original sweet pea can be grown still, called 'Cupani' after the Sicilian monk who sent the seed. Cupani flowers are tiny, dark blue and dark pinkish-purple, with only one or two flowers on short stems, and the scent from these tiny, truly pea-like flowers is so sweet and strong that a posy of ten stems will scent a whole house in warm weather. Cupani was soon bred into different varieties, including the heritage pink-and-white 'Painted Lady', which can also still be found as a seed to grow today.

Heritage varieties are not brilliant to grow for commercial production, but gorgeous to grow for yourself. They don't last particularly well as a cut flower – three or four days maximum. Small bunches at a farmers' market will sell out as soon as you put them out, so long as you don't grossly overprice them.

Grandiflora varieties

At the end of the nineteenth century the grandiflora sweet peas were developed: longer-stemmed and bigger-bloomed than the heritage varieties, with three or four flowers per stem rather than one or two, but still smart and simple compared with the outrageousness that was soon to come. The grandiflora series of sweet peas now numbers hundreds of varieties, all highly scented and easier to cut than the tiny, original Cupani, because those longer stems are stronger and bruise less easily, and can be conditioned in ordinary buckets rather than in shallow, fiddlier jars or vases.

I find that the grandifloras have a slightly longer vase life than the fleeting heritage varieties. There are a great many to choose from: trawl the seed suppliers, and always experiment.

Spencer varieties

In 1900 Silas Cole, a gardener at Althorp (Princess Diana's family home) developed the first of what became known as the Spencer varieties. These are the French knickers of the sweet pea world. Suddenly, what had been a rather refined, restrained, if highly scented line-up was invaded by outrageous show-offs. Tall, strong, less swooningly scented, the ruffles of a big Spencer sweet pea can make flower heads as big as roses. Cole's contemporary William Unwin developed similar ruffle-edged sweet peas, demand for which was so great that they formed the basis for that monster of seed supply Unwins Seeds, still blooming as a business today.

Fully out, these outrageous girls, with up to five wildly ruffled flowers per stem, are as beautiful as anything you can grow in the garden. Less scented than their grandiflora parents or heritage variety grandparents, their perfume still fills a warm

Spencer sweet pea 'Charlie's Angel' – already fabulous and still flowers to open.

The pure white sweet pea 'Royal Wedding' has a wonderful scent.

room on a sunny day. Put them near an open window so that the breeze blows the scent right through the house.

For cut-flower production, the Spencers command the highest price, because of stem length and their showy heads. Think of what kind of market you're planning to sell to: heritage varieties take as much time to grow and cut as the show-off Spencers, and yet a posy of them will be less than a third of the size of the same number of Spencers tied. But I still say that heritage varieties in tiny bunches for farmers' market stalls would sell just as well as the more showy Spencers, simply because of the scent and their heart-string-tugging flower-fairy-like sweetness. And remember, you are the person who will be selling your crop, so grow the sweet peas that touch your heart most, because they'll be the easiest for you to sell.

Choosing colours

Choose your sweet pea colours carefully and they can be an integral part of your 'look'. If you're selling by the stem to florists or wholesalers, remember that the pale, weddingy colours will fetch the best price per stem. Ask them which colours they'd like and work with them: they know their end customer – the more *they* can sell, the more *you* can sell.

Brights and darks are fantastic too, especially if you're growing to have a late-season crop and plan to mix them with some jewel-coloured dahlias. Mixed bunches will pay less per stem than single colours.

If growing for gate sales, you should still consider growing, say, three separate colours of sweet peas. That way you're giving your customer three choices. At a good price they might buy all three bunches. If you give them lots of bunches of a

Think of what kind of market you're planning to sell to: heritage varieties take as much time to grow and cut as the show-off Spencers, and yet a posy of them will be less than a third of the size of the same number of Spencers tied.

mix of colours, your barrow looks as though it's filled with repeats of the same thing and your customer will only see one option to buy. It makes commercial sense to give your customer the opportunity to spend more if they'd like to: if they can't decide between the pink and the lilac, they might well buy both.

Mixed-coloured sweet peas are an acceptable option if you only have limited space, but you can't control the numbers of stems of a particular colour that you'll get, and if a customer asks whether you can supply a lot of a particular colour on a particular day, a) it'll take longer picking them if you're fiddling about between other colours to find the one you want, and b) you won't be able to calculate in advance roughly how many you'll have.

My top five sweet peas to grow for cut flowers

These sweet peas come top of my list, but, as ever, your line-up may be very different – your taste in colour, scent and showiness much more refined than my rather vulgar love of the most outrageous of the sweet pea flowers. . .

The dark red 'Beaujolais' goes beautifully with jewel-like rich-coloured dahlias in bouquets.

'Beaujolais'

This flower really is the colour of a ripe red wine. It's not especially scented, and it can be easily damaged by rain and wind, but we always grow it because of its fabulous colour, especially in a later crop, so we can mix it with rich-coloured dahlias.

'Betty Maiden'

White with a blue picotee edge, large-flowered and beautifully ruffled, Betty Maiden is like a

Dior evening gown from the 1950s. It's much in demand by brides, so is beloved of wholesalers.

'Charlie's Angel'

This is the strongest-stemmed of the sweet peas we grow. It's a lovely lilac / pale blue (when sweet peas are called blue the blue is always really quite lilac) and a prolific producer.

'Mollie Rilstone'

This girl is cream with a pale raspberry (but not at all sugary pink) edging to the ruffles on her generous flowers. I would grow Betty Maiden and Mollie Rilstone before all others.

'Royal Wedding'

Pure white, highly scented, great for weddings. Royal Wedding doesn't have such huge flower heads but does have great scent – and for a bride who wants white white white, it's no good offering her white with a touch of blue or pink.

Cutting and conditioning sweet peas

With their delicate petals and unmatched scent, sweet peas are designed to be over quickly. Grow them for their fleeting beauty, love the way they scent your whole house with their perfume, and compost them when they're finished. They're sweet peas: they will have gained plenty of sisters while they were on your or your customer's kitchen table. Cut more fresh ones; compost them when they're over. Flowers are not supposed to last for ever.

That said, as a flower farmer you are of course growing for sale, and the more life you can coax out of your cut stems, the more value they will have for you and your customer. With the right treatment your cut sweet peas should last between 5 and 7 days. If you choose not to treat them with flower-preserving silver nitrate, they may last a day or two less than treated flowers, but they will keep their glorious scent – and they will grow in the vase, changing the shape and story of the bouquet they're in.

Cut sweet peas when the first flower is coming out and the others are still in bud. The rest of the flowers will come out in the vase. Cut them early in the morning, when they've had a good drink through the night and the water is naturally high up in their stems. If cutting in the evening, wait until the sun is no longer warm on the back of your neck. As with any flowers, always cut them direct into water (see Chapter 12).

Without crowding them, which might bruise their petals, especially if they're wet when you cut them, put the flowers neck-deep in water and let them have a long drink in a cool, dark place, for a minimum of 4 hours; ideally 12.

When arranging sweet peas, snip their stems once more to prevent the stem ends sealing over and to keep the water-absorbing cellulose cell structure open for the water to rise up. Change the flower water every day or so, especially if it's beginning to look cloudy. Snip the flowers' stems when you do so, to keep the water's access fresh. In hot weather, putting a teeny drop of bleach or white vinegar in the flower water will keep it clean: bacteria in flower water will kill flowers quickly.

If you want to use a sugar solution for your sweet peas, then 40 grams of sugar dissolved in a litre of water (1½oz in 1¾ pints) makes a good home-made flower preservative: a mixture which a great many growers swear by. We don't use anything like this at Common Farm, and our sweet peas last up to a week.

You may find that a wary local trader looks askance at your sweet peas and says, "But they never last." Put your just-coming-out sweet peas in a pretty jug and offer them as a gift. Ask the trader to keep them in their shop for a week, then go back at the end of the week to see how they did. Florists often think that if flowers come to them without the stamp of a Dutch auction on them, then they can't be good quality. It's up to you, the grower, to prove the opposite.

Bouquet combinations with other flowers

Rather than ask yourself, "What grows well with sweet peas?", perhaps you should ask instead what job the sweet pea does in a bouquet of flowers. Look at its ruffled edges; the crisp gauze of its texture. Is it big enough to be the centre of attention in a bouquet? What would frame a sweet pea?

The answer is to think about *texture*. For sweet peas to be the centre of attention you have to frame them with something lighter: ammi, moon carrot, orlaya. Then give them smaller buttons, as though you were adding buttons to a silk crêpe dress: cornflowers do this well.

Clary sage has a similarly tissue-papery feel to the coloured leaves climbing its stem, and makes an attractive architectural spike in a bouquet of sweet peas.

Sweet pea posies mixed with ammi, sweet rocket and alchemilla.

A bride's bouquet with kind old rose 'Madame Alfred Carrière', sweet peas 'King's High Scent' and 'Betty Maiden', and alchemilla.

Cerinthe, with its silvery leaf and subtle, almost hidden, purple flower, makes a good frame for a bouquet of sweet peas.

Similarly, that old staple alchemilla, with its cloud of acid-green flowers, will froth around the matt fabric which is a sweet pea.

If sweet peas are to play a supporting role in a bouquet, then they can be used to edge any sort of bigger, showier flower: roses, peonies, dahlias can all be given a crêpe de Chine ruffle with sweet pea edging.

This approach to working out what to grow in conjunction with a particular flower works well with all flowers. It's worth doing, not only if you plan to use your flowers in floristry yourself but also if you're building a market selling to local florists: the more of a package you can supply, the better a service you're supplying.

Five top tips for growing sweet peas for cutting

❋ Plant sweet peas 30cm (12") apart in humus-rich soil. They are hungry beasts, and they don't like their feet to dry out.

❋ Stake them well and give them netting to climb. If you grow them up a teepee, you risk losing flowers and therefore missing developing seed pods in the jungle centre of the structure.

❋ Choose sweet peas for your purpose: heritage varieties are heart-string-tugging teeny and have most of the scent; grandifloras are bigger, still highly scented; Spencers are the vulgar, showy girls in the mix (and my favourites).

❋ Think about who you're growing sweet peas for – weddings, markets, bouquets . . . and choose your colours accordingly.

A few words from Ursula Cholmeley

Ursula Cholmeley has spent over a decade recreating a lost 12-acre garden, Easton Walled Gardens in Lincolnshire, which is a great place to see sweet peas in full glory.

At Easton there is a beautiful cut-flower patch next to the tea room, so you can sit and enjoy scones and sweet peas together. They grow so many sweet peas here that they have special 'sweet pea weeks' when the flowers are especially abundant, and visitors may cut them to take home.

Here you can see in the flesh, so to speak, the difference between grandiflora and Spencer, and smell the scent of the heritage varieties.

"Our great love affair with the sweet pea came when we created a cut-flower garden, 'The Pickery'. This gave us a chance to do some real horticulture while the rest of the garden restoration was a mess. We grew a few sweet peas for arranging, and got carried away. Now we grow nearly 100 varieties from two sowings each year, one in early November and one in the spring.

The Pickery is based on six beds, of which about half are given over to sweet peas (there are more in the cottage and vegetable gardens, and in the rose meadow and borders.) Our earliest mistake was to be too mean with the size of the beds. The sweet pea roots couldn't spread out, and small beds meant lots of edging and maintenance. Now we have big wide beds with lots of space for each row.

The beds are rotated so sweet peas are not grown in the same place every year. This means that it is sometimes necessary to put a temporary path down the middle of the bed. The sweet peas seem perfectly happy with this arrangement, and their roots can wander under the barked paths.

Here are my three hot tips for sweet pea growing:
* Sweet peas like a really good root run. Dig your ground over and add plenty of well-rotted manure on a light soil. On heavy clay you may need to add sand or gravel to open up the structure. Either way, make sure you dig the ground over well.
* If you are unable to do this, grow the old grandiflora varieties in a container. They have smaller flowers but make bushier plants and are less demanding than modern varieties. Water regularly with a liquid feed.
* Sow in early May for late flowers in September, when the weather can be especially favourable for flower production. In autumn you will still be picking sweet peas when other hardy annuals have had enough.

And a few of the Easton favourites:
* For sophistication and charm: 'Mollie Rilstone'.
* For being good-natured and a good doer: 'Gwendoline'.
* For scent: 'Matucana'.
* For depth of colour: 'Our Harry'.
* For drama: 'Linda Carole'.
* For nostalgia: Easton Walled Gardens Heritage Mix (includes varieties such as 'Lord Nelson', 'King Edward VII' and 'Painted Lady')."

Chapter ten

Herbs

Rosemary for remembrance; lavender to calm a nervous bride. . . Mint for a sharp contrast to the sweetness of a rose's scent. . . The heady aniseed depth of late-summer bronze fennel. . . Herbs are often forgotten in a cut-flower scheme, and yet I know that florists love them, so don't overlook these valuable plants – they will complement the rest of your crop.

Pineapple mint in garlanding for a wedding.

For fabulous foliage fillers and unexpected scent sensations, not to mention medicinal value, I recommend you create a herb patch in your cut-flower garden. Herbs are difficult to get hold of for florists, and so they will often be happy to buy from a local grower if they can't rely on their wholesaler for these. Herbs don't generally travel well out of water, and are often easily bruised, which makes them expensive to import.

Ask a bride whether she'd like fresh herbs as ingredients in her bouquet and you'll see by her reaction that it is a good idea to grow them.

If you're growing in a confined space and you want to make every stem you produce count financially, then herbs are a good crop. Imagine the lovely furry apple mint in a bouquet with garden roses; the bicoloured pineapple mint as a striking filler in white wedding flowers; the delicate steep white spikes of their cousin melissa's flower contrasting with sweet peas. . .

Think too about herbs whose flowers you don't usually see because the leaves are cut too young for the plant to produce flowers or seed: parsley, sage, marjoram or thyme for a buttonhole.

Grow your herbs in a relatively protected environment. I have found rosemary less hardy than lavender: though both overwinter happily enough, however bitter the frost, rosemary will die if a late frost bites hard once the sap is rising again.

We don't grow masses of lavender or rosemary. It's impractical to expect a great crop of these Mediterranean natives from our thick clay. I keep both in pots around the yard: well drained; relatively frost-protected; handy for buttonholes, brides' bouquets, etc. Whereas mint I grow in the field. If your conditions are different, you may find you do the opposite.

The subtle acid-yellow lace of parsley in flower is useful for edging stronger accent blooms.

Early in the season our herb patch looks quite tidy.

Which herbs to grow?

Consider the following list with a little circumspection. If space is limited, you may simply decide to have a low hedge of herbs around your cut-flower patch. Angelica, for example, is a space-greedy plant, and so might be one to grow if you're planning a bigger herb area. A hedge of lavender will give you a low windbreak as well as flowers for cutting. Think about the options and choose what will suit you.

Recommended herbs

Even in a small cut-flower patch I particularly recommend you start at least with a little mint, some rosemary and a lavender or two. You can expand your collection as your business and growing customer base demands.

Angelica

With giant round heads of tiny white or pale pink flowers on green or dark red stalks, this perennial herb is an impressive flower to sell by the stem, and extremely useful for large arrangements at events or weddings. It smells of gin and tonic when cut. Collect seed in autumn and sow straight away, as it germinates best when fresh.

Bronze fennel

Cousin to the dill, bronze fennel is a short-lived perennial and freely self-seeds for us. It has a more mustardy-coloured flower than dill, on bronze stems. The aniseed scent is gorgeous to me, although some people dislike it.

Warning: keep dill and fennel apart, or your dill will become fennel!

Bronze fennel makes a wonderful filler for flower arrangements, though some people don't like the aniseed scent.

Catmint

This is an excellent, useful perennial to use for foliage as well as flowers. Cut back after the first June flowering to encourage a second flush in the autumn.

Comfrey

While not one to grow as a cut flower, this perennial is necessary for making comfrey tea to feed your garden in the second half of the summer, or to chop up as a mulch for your beds. It also feeds the bees.

Coriander

An annual to grow every year. Let some of your coriander go to seed, for gorgeous umbellifer heads of soft white lacy flowers.

Dill

This bright yellow lace-capped annual flower is extremely useful for floristry. Check seed suppliers for 'flower arranger's dill' – a slightly bigger, taller variety than common-or-garden dill, though I like the small heads of the ordinary flowers too.

Feverfew

The tiny daisy heads of this annual chrysanthemum are incredibly useful in posy making, I find, though I've had clients who've turned their noses up at this smiling little plant as a 'weed'. It will seed itself everywhere, but you soon become adept at recognizing seedlings and either rationalizing them into one place or composting any you don't need. I think it so useful that I plant it afresh each year, for fear there won't be enough self-seeded examples about my plot.

Hyssop

Not only good for keeping moths out of your woollens, flies out of your kitchen, and growing a short-lived perennial hedge as a windbreak in your plot, hyssop is lovely cut: a true blue for late-summer bunches.

Lavender

If you have the ground for it (free-draining), grow large quantities of lavender. There's a great market for it for everything from lavender bags to herb teas, from cooking to soap, from floristry to essential-oil making. People scatter grains of lavender on paths into reception marquees or coming out of church, so that when guests tread on them the scent is released into the air. Lavender is calming to a nervous bride, a spot of real blue in a bouquet or buttonhole, and dries beautifully for remembering the day. . .

The French cropping lavender you see stretching in fields to the horizon in Provence is one of the taller varieties, called 'Grosso'. You should think about whether you're growing lavender for height

Ask a bride whether she'd like fresh herbs as ingredients in her bouquet and you'll see by her reaction that it is a good idea to grow them.

or not. We don't bother particularly for height: we're growing for posies, for brides' bouquets and for buttonholes. You may be growing to sell in bunches at a farmers' market or to your local florist, in which case a taller variety may be of more use to you.

Lemon verbena

A perennial, shrubby plant, lemon verbena is very tender, so bring it in under cover for the winter if you're gardening in an exposed spot, but the incredible lemon scent is amazing in bouquets. It won't last a minute out of water, so don't ask it to stand for long in flower-foam-based arrangements. As a foliage plant it has a short season, but it repays the time and space you give it a hundredfold with that stunning scent and its attractive, pointed, acid-green leaves.

Melissa

A close cousin to the mint, melissa has lovely tall spikes of white flowers in the summer. Its scent is subtly lemony, a delicious addition to a cut-flower mix. We grow it in dry shade, but it will tolerate damper conditions.

Mint

The scent of mint in a cut-flower arrangement is unexpectedly delicious – like a sprig of mint on a ball of ice cream, with a chocolate pudding, or with raspberries.

Mint will bully its way about a bed where it's planted, but its roots are shallow, so plant in blocks of its own kind and separate with shallow barriers across and just under the soil surface.

But beware: mint roots will work their way through a piece of wood, finding their way through the smallest gap, so do lift wood barriers at the end of the season and check for invasion. You could also use old slates, stone or plastic as a barrier, but still check them at the end of the season, because your mints will try to mingle, and if they do you'll end up with one generic green mint and lose all your gorgeous varieties.

In a small patch you could cut the bottoms off plant pots and grow your mint inside the space encircled by the plant pot. Keep the pot top an inch or so proud of the soil in your bed and the shallow roots won't be able to spread.

These three are fine mints to try:
- apple mint: has soft, furry leaves
- pineapple mint: an attractive variegated colour
- buddleja mint: for tall spikes of flowers.

Oregano

Very useful as a little scented purple cushion in bouquets, this perennial herb likes dry conditions and will self-seed about in a path or a yard. We grow it in pots, on a path and in the herb garden, and it flowers at different times and for longer/shorter periods, depending on the summer we're having. I wouldn't be without it for its insect-drawing abilities either. It's very similar to marjoram, and the two are often confused, though marjoram has a slightly sweeter, gentler flavour.

Parsley

Sow flat-leafed parsley as a biennial and you'll have lovely tall greeny-yellow lace-capped flower heads in late spring the following year.

Rosemary

For funerals, weddings (it is traditionally associated with remembrance, and so a bride will often ask for a sprig in her bouquet in memory of a beloved grandparent), for winter greenery and Christmas wreathing – many customers have a special fondness for rosemary, so it's worth growing. 'Miss Jessopp's Upright' is the tallest.

Sage

This lovely herb, with all its variety of colours from purple to silver, makes wonderful foliage for cut flowers. Grow the variety 'Hot Lips' for a burst of blinding pink flower spike to add to dahlia bouquets at the end of the summer, and the classic 'Victoriana' for pure blue spires, which go beautifully with the rich yellow of a sunflower and the mustard of bronze fennel. We lift our sages and overwinter them in the tunnel for fear of frost, as they can be tender, and always grow fresh new plants every spring. 'Hot Lips' cuttings will root in a glass of water.

The annual clary sage, in all its pink, white and blue varieties, offers summer-long delight – see Chapter 2, page 44.

We keep our herbs in a drier, higher, warmer bed than the rest of our crops, next to the house – our field is too damp for most of them.

Strawberry

Not strictly a herb, but the flowers of the wild strawberry plant are beautiful in bouquets, though very fragile, so better for a bride than for a bouquet designed to last a week. Wild strawberry varieties shoot longer stems than the fatter, domesticated strawberries. I use both in floristry.

Thyme

Thyme is good for buttonholes, brides' bouquets and jam-jar-style wedding arrangements, but isn't really tall enough for everyday bouquet work.

Cutting and conditioning herbs

Be careful with soft-stemmed herbs. They need a good overnight in water to really fill up to their necks. Basil (beautiful tall spikes of flowers) will cut and stand in water, but don't expect it to last a minute in a bouquet being carried by a bride or in a flower-foam-based arrangement. And take as much care with woody-stemmed herbs. Cutting into the wood can make it difficult for the water to go up the stem, but you may need to do so in order to get the stem length you want. Even if the stems are very thin, it's worth splitting them an inch to increase the surface area from which the flower head or leaf tops can drink.

If herbs wilt despite good conditioning, you can try searing the stems with an inch of boiling water for 30 seconds, before filling the same container with cold water and leaving the herbs to have a good drink before using them in floristry. See Chapter 12, page 189, for more on searing.

Always re-cut the stems of herbs after making a bouquet, or any arrangement in which the material has been out of water for even the shortest time.

A strawberry flower in a groom's buttonhole.

Five tips for growing herbs for cutting

✽ Grow what grows well in your soil. Mediterranean herbs don't like a damp, wet clay, but mint doesn't mind it.

✽ Don't try to grow woody-stemmed herbs from seed: they're often easier to grow by rooting cuttings. Lavender and rosemary root readily from cuttings. Mint will send out roots from a cut stem in a glass of water. Grow annual herbs from seed.

✽ Be wary of the spreading mint: you'll never get rid of it if you plant carelessly.

✽ Lift and overwinter tender herbs in your polytunnel. Take cuttings mid- to late summer to replace any plants you may lose to frost.

✽ Many herbs have very useful properties: some are antibacterial; others are pest deterrents. Grow some mint in your tunnel and dotted about your other flowering beds to fight mould. Underplant roses with sage, and it will help keep the aphids off, because they don't like the smell.

Chapter eleven
Wildflowers

You may have heard that wildflowers don't last as cut flowers. Not true! Include some of these beautiful flowers in your collection and you'll have the satisfaction of knowing that you're contributing to the regeneration of our wildflower seed bank. Plus you'll be giving a helping hand to the wildlife, who will benefit you in return.

The point of growing wildflowers for cut flowers is twofold: yes, they do make marvellous cut flowers, but if we all grow a patch of wildflowers, however small, we are also growing a little environment in which native wildlife can flourish.

A tithe to nature

We began Common Farm Flowers because it was a compromise between our need to earn a living and our desire to make an ecological sanctuary for birds, bees, hedgehogs and ladybirds. . . How could we afford to keep our little patch a haven for wildlife and plants *and* pay the bills?

When we found the place, Fabrizio had a strong vision of what he called 'hedge world' (since we moved here we've planted just under half a mile of native hedging in our 7-acre plot). He saw our meadows dotted with shallow ponds, a corner for honey bees, plenty of shelter and forage for grass

You might think about keeping your own bees – or invite a member of your local beekeeping association to take advantage of your forage-rich plot for theirs.

snakes, slow-worms, toads, newts, bats, butterflies, dragonflies (the air is alive with dragonflies here in the summer), and so on. But how were we going to pay for it all? Land that doesn't work for a living is not a treat we can afford.

We have butterflies in profusion here at Common Farm. This tortoiseshell is posing tastefully with echinacea.

When it came to us, the cut-flower farm idea seemed so obvious that we were amazed it had taken us so long to think of it.

Eco-structure

The structure of our flower farm is that of rows of beds interspersed with protected areas of intensively created meadow, connected to one another by 'motorways' of mown paths, blocks of orchard, patches of oak, native hedges planted to protect our flowers from the wind, heaps of horticultural detritus tucked into corners where the hedgehogs can nest, heaps of steaming-hot grass cuttings topped off with corrugated iron plates under which the grass snakes can knot themselves about one another in the warm.

We use no commercially prepared chemical weedkiller or pest control. We simply feed our earth and give shelter to the biological-pest-control army. Ladybirds eat our aphids. Birds eat our snails. Toads and hedgehogs eat our slugs. Snakes and birds of prey eat our mice. Our ladybirds overwinter in our euphorbia collection. We've put up any number of bird boxes on the north faces of trees (put them on the south face of trees and the hatchlings will cook on a hot day). We get our slow-worms delivered to us in the municipal green-waste compost we buy by the lorry-load!

Fabrizio always says, "Look after the invertebrates and the rest of the food chain will look after itself." And so we grow wildflowers to feed the butterflies, poplar to give the bees propolis (with which they glue their hives together), willow and hazel and ivy on which the bees can feed in deep winter. . .

Avoid chemical sprays and encourage wildlife.

The food chain is clear here at Common Farm: watching the barn owl hunt on an early June morning as I cut sweet peas, I know he wouldn't be here if there weren't mice, voles and shrews, and they wouldn't be here if there weren't slugs and worms and beetles, and they wouldn't be here if there weren't. . . The owl is amazing to watch. He flies surprisingly low, his wide wings occasionally pumping the air with a powerful *wumph*. He is confident, and seems to ignore me marching up and down the flower rows with my trolley and buckets and snipping scissors. He'll perch on the tall stakes dotted about the place, watching. The tall stakes were put in to protect the young trees we planted (birds of prey are heavy; young trees fragile). We occasionally have a pair of kestrels hunting the meadows here, but when the barn owl moves in, the kestrels are nowhere to be seen.

Fabrizio always says, "Look after the invertebrates and the rest of the food chain will look after itself."

Allow ivy to scramble about – it will provide winter shelter for invertebrates, fodder for bees, and foliage for your cut-flower arrangements.

Red-soldier-bug romance on a wild carrot flower. These bugs are great polinators, the adults feed on aphids, and the larvae prey on slugs and snails – love them!

Large-scale or small-scale

You don't need a large area to create a haven for wildlife: in a small patch, you just need to think proportionally. Whatever the size of your plot, you'll still need a compost heap; you can make a bug-house for beetles and woodlice from a brick or log, flat against a wall; solitary bees will nest in holes in masonry. Even the smallest pond will provide water and cover. Make cover and the wildlife will use it.

Do grow a patch of ordinary stinging nettles (or rather, allow them to grow), not only to use in your nitrogenous home-made nettle tea but also to feed the butterflies. Peacock and tortoiseshell butterflies breed and feed on stinging nettles. So, in the spirit of keeping up your tithe to nature, welcome a patch of stinging nettles to your land.

The biological-pest-control army

There is a balance which comes quite quickly to a piece of land gardened organically, if you hold your breath and have faith that it will come. Every year I narrow my eyes at the first hatch of greenfly on the roses, and soon afterwards the ladybird larvae appear, munching their way through them. It may be that your garden needs a little help, though, and there are ways of encouraging pest controls into your patch in a more strategic way than just calling for them into the wind.

Ladybirds

Ladybirds are your aphid-control army. Look on-line for purveyors not only of nematodes to eat off slugs but ladybird larvae with which to populate your polytunnel – valuable if you need a bit of extra help to get you started.

Aphid control if the ladybirds are not hatched yet

Liquid soap can be heavily diluted and the mixture used to douse the aphids if your greenhouse or polytunnel is hot in the winter, or they're breeding in the early spring before the ladybird larvae hatch (always about a fortnight of scary time.) Spritz this mix on to aphid-infected areas two or three times over several days and the aphids will die off. Never spray leaves with soap – or feed or any kind of solution other than water – when the sun is hot on the plants, as their leaves will be singed. Wait until early morning or late afternoon, or until the weather's clouded over.

Toads

These are glorious creatures: pick one up and it will pee in disgust all over your hands, so it's all you can do not to chuck it with a 'yeurgh!' into the hedge. They are also enthusiastic slug-eaters. Keep your toads to hand, keep them in your cut-flower patch, and they will do good work for you.

I have to remember not to give toads shelter in the hedge when I find them as I clear beds at the end of the summer – because by the end of summer the toad population is made up of large beasts which wouldn't be able to get back to the slug-fest available in my cut-flower patches, as they're too big to get through the rabbit-proof fence. So I keep heaps of gardening rubbish, what some people might describe as bad compost heaps, in which they can shelter (along with the hedgehogs, whom I seldom see but sometimes hear rustling about in the undergrowth).

Keep your toads to hand, keep them in your cut-flower patch, and they will do good work for you.

Grass snakes

Eaters of slugs and mice alike, grass snakes are creatures to encourage into your patch. No, grass snakes are not frightening: they neither bite nor are poisonous. They are gentle, shy, sun-loving creatures with soft, dry skins and a devious way of playing dead if you pick them up.

I've heard of lucky flower growers who have grass snakes set up home in their polytunnels (polytunnels are favourite nesting patches for bad, seed-and-seedling-thieving mice). We have quite a collection of grass snakes, including Old One Eye, a fat hen snake who must be six years old. She lost an eye, to a mower's blade we think,

Young grass snakes: the one on the left is looking up to see if you've noticed he's been playing dead like his brother on the right.

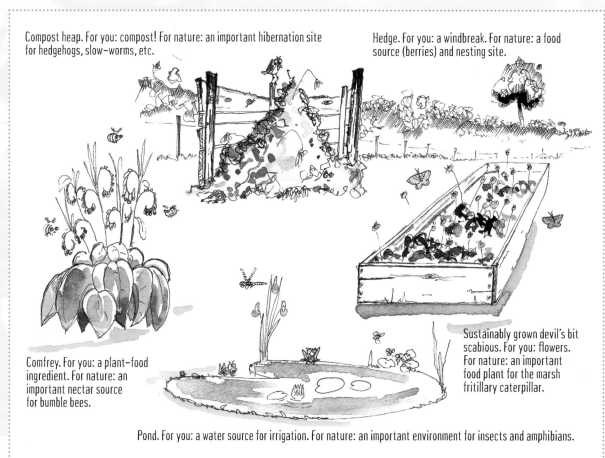

Compost heap. For you: compost! For nature: an important hibernation site for hedgehogs, slow-worms, etc.

Hedge. For you: a windbreak. For nature: a food source (berries) and nesting site.

Comfrey. For you: a plant-food ingredient. For nature: an important nectar source for bumble bees.

Sustainably grown devil's bit scabious. For you: flowers. For nature: an important food plant for the marsh fritillary caterpillar.

Pond. For you: a water source for irrigation. For nature: an important environment for insects and amphibians.

The wildflower garden. Every element benefits both you and the wildlife.

but she appears on hot summer days to bathe in the sun's rays on the stone Fabrizio filled the ditch with so that we could push a barrow over it, down between the Gilly orchard and the wild-flower meadow. Grass snakes, when caught, play dead quite effectively, though their habit of looking up to see if they've taken you in with their trick slightly gives the game away.

Establishing a meadow

You will probably end up growing some wild-flowers in a bed and some in a meadow or orchard. To split your wildflowers like this means you'll have two different kinds of crops. Flowers forced to fight with grass for space and nutrition are more delicate, finer and less numerous. Wildflowers grown in beds use a lot of space for their flowers, but give you stronger, taller flowering stems.

We have *never* successfully established a wild-flower meadow through the 'scrape and sow' system of taking off the topsoil from a section of ground, sowing wildflowers, and leaving them to establish. Our meadows were sown in the past with strong rye grasses for feeding cattle. The earth in our meadow is not easy for a wildflower seed to find purchase and germinate in, because the clay is so thick. So we tend to grow our wildflower plants from seed in trays and plant out seedlings when they're big enough. Those planted in beds are protected from rabbits by the careful fencing of all our cut-flower plots. Those planted in the meadows between need protection: from January onwards we pop blue plastic mushroom trays over our plantlets and hook the trays down with a little twig on either side. This puts the rabbits off and allows the plants to start producing leaves and develop flowering shoots, ready for when the growing season is under way enough for them to escape the rabbits, deer, mice, voles. . .

Yellow rattle is a predatory wildflower, the roots of which eat the roots of grass. This means it makes space for other wildflowers to colonize an area of meadow, which is why it is much sought-after as a starter for making wildflower meadows. We've used kilos of yellow rattle seed over the past ten years, and are slowly establishing patches. It's been an expensive process, but once the patches take hold they spread fast, cutting down on the success of the meadow grasses and making it easier for other wildflowers, planted out as seedlings, to establish. Yellow rattle is expensive because you can't plant it in seed trays to germinate. You have to sow it direct and hope it'll take. And you need a lot of seed when direct-sowing, because the percentage of seed germination and ultimate success is small.

Meadows are not easy to establish, which is why it's such a tragedy that in the UK we've lost 97 per cent of them since the Second World War. But with patience, care and a little cunning, we can bring at least some of them back. Your cut-flower patch can play a vital role in rebuilding our native seed bank and encouraging wildflowers to recolonize corners, strips, edges, or even whole acres of meadow.

Which wild plants to grow?

You may not have an enormous amount of space for wildness in your patch, but think about what might grow well for you, and whether a little wildness in your product might set you apart from your cut-flower-growing neighbour, as well as give you an edge on the imported flower market.

Wild hedge plants to grow for cutting

A hedge of native shrubs to use as cut-flower foliage might include wild dogwood, guelder rose, crab apples and hawthorn – all valuable at different stages of the season.

Avoid wild roses because they'll kill a hedge if not kept under control, and, while their flowers are exquisite, they don't last well in water for more than a day, and they have such vicious thorns that you'll rip yourself to shreds no matter how armour-plated you consider your gauntlets to be. Equally, think hard before planting old man's beard (wild clematis) as part of your hedge, as it will strangle an establishing hedge quickly: you'd do better to keep an eye out for good sources of it elsewhere, make friends with the farmers who have it lacing their hedges, and get permission to cut it for yourself. Farmers are often happy for florists to cut their old man's beard because of the plant's hedge-strangling habit, but be sure to ask – I wouldn't want to be accused of encouraging anyone to trespass!

Spindle, rowan, holly, beech and oak

All these trees are not only good for wildlife but are also useful for use in cut-flower arrangements. Acorns make fantastic highlights in autumn

An autumn bridal bouquet with old man's beard and wild carrot in the mix.

bouquets and buttonholes, or scattered down the tables at a wedding reception. Have these trees as a haven for wildlife and the wildlife won't mind sharing their environment with you. Holly is very slow-growing, so one for people in for the long haul or who've inherited good stock from previous owners of their land.

Blackberry

Another hedgerow bully, with horrible flesh-ripping thorns, but the blackberry does make a gorgeous addition to an autumn bouquet. There's something about a shining button of blackberry in a bunch of flowers that always makes the recipient smile. Cut blackberries when the fruit is still greenish-red (a ripe blackberry will simply squish itself all over the other flowers and make a not-

very-charming mess). Of course the berries will ripen in a bouquet over time.

Dogwood

For the purple-edged leaf bud in early spring, for flat cushions of tiny white flowers in May, and for aubergine-coloured leaves and little bunches of black berries in October, the wild dogwood makes enormously useful foliage and a good, fast-growing windbreak.

Elder

The foliage of the wild elder will condition well in early spring or autumn, but in between it can be unreliable and wilt in the heat. I've never been able to condition the flowers reliably, however.

Guelder rose

A top favourite for cut flowers. In spring you have lovely fresh, super-green foliage and lace-cap flowers – especially useful for big wedding arrangements, but lovely in bouquets with the first roses too – and then in autumn there are bunches of bright-red berries with red-edged leaves. Guelder rose is susceptible to attack by viburnum beetle, so plant it sparingly in mixed hedges and the beetle might not find it.

Hawthorn

The Beltane flower; the May flower. In far off, pre-Christian times, circlets of hawthorn blossom were made for May Day gatherings at which marriages were celebrated. More recently (well, between six and eight hundred years ago), it's been tainted by the fact that blackthorn thickets were planted over plague pits to protect the bodies from marauding packs of dogs, and you'll often find people who won't allow it in the house. It's an understandable mistake to make, though the blackthorn flowers a month or so earlier than the hawthorn, and has even more vicious thorns than the hawthorn.

When blackthorn flowers, in March or April, it often comes with a flash of cold weather, which is traditionally described as a blackthorn winter. But when hawthorn flowers, in late April or May, it's traditionally time to take off your winter coat: "Ne'er cast a clout till the May is out" means "Don't take your winter layers off till the hawthorn is flowering."

Hawthorn blossom stands well in water, though the leaves, which open first, tend to wilt, so if you're going to use hawthorn, do strip the foliage (a fiddly job) before using the blossom in your cut-flower arrangements. The berries are useful in autumn arrangements too, and at that time of year people seem to be less superstitious about the plant. Equally, sprays of blackthorn berries – sloes – are great in autumn compositions and wreaths, and people seem to forget that they are the fruit of the blackthorn (so scary in spring; so gin-associated in autumn).

Hazel

Hazel will give you catkins in deep midwinter: here we cut them from January through to March, depending on the winter we're having. The catkins will shed pollen when brought into the house and allowed to come out fully, but you can spray them with a little hair-spray to stop this.

Willow

Crack willow, corkscrew willow, blinding-coloured cultivars of willow . . . all of them will give you gorgeous spikes of pussy willow at some point between February and April to cut and sell. If you have damp land, then plant wil-

When blackthorn flowers, in March or April, it often comes with a flash of cold weather, which is traditionally described as a blackthorn winter. When hawthorn flowers, in late April or May, it's traditionally time to take off your winter coat: "Ne'er cast a clout till the May is out" means "Don't take your winter layers off till the hawthorn is flowering."

low – you can simply strike a metre-long (three-foot) stake of fresh-cut willow and it will root. You can also use it to make wreaths at Christmas and throughout the spring for weddings, Valentine's Day, Easter and so on. Willow has ancient associations with Wicca magic – inspired, I'm sure, by the fact that it's so early-flowering, promising the magic of spring. Bunches of pussy willow, or pussy willow in your posies, will make every customer smile and reach for them.

Cut willow early and bring indoors to force in a little water: this way you'll have pussy willow and spring leaves earlier than if you wait, and you can stagger your saleable crop.

Wildflowers to grow for cutting

So, grow wildflowers for the benefit of your environment – but if you're giving them space in your patch, you want them to work as cut flowers too. Here's a short list of those that we've found good for cutting here at Common Farm.

Bluebells

This lovely bulb naturalizes in woodland and hedgerows. You *can* cut bluebells to work as cut flowers, and yes they *will* last. But for goodness' sake cut them in the cool of the morning and straight into water. Cut when the bottom few flowers on the spike are just opening, not when the whole flower is out. This way, you have bluebells for your flower arrangements.

Remember, the urge may be strong, but you *must not* cut even one stem from a wood of thousands of bluebells carpeting the ground unless that

A wildflower posy for a spring wedding: including bluebells, fritillaries, cowslips, red campion, cow parsley and forget-me-nots.

CHAPTER ELEVEN Wildflowers 175

A wildflower posy with buttercups, ox-eye daisies and cow parsley, plus physocarpus for greenery.

wood belongs to you or you have clear permission from the owner.

You can order bluebells in the green (dug up in clumps already shooting) in January, and as bulbs for autumn planting. They may take a while to establish, and unless you really have serious itchy-scissor syndrome, and also have a great many of them, you may find it difficult to cut them when you could just lie down in them and indulge a dream of blue.

A note on species purity: when buying bluebell bulbs, be sure to order the native wildflower *Hyacinthoides non-scripta* rather than the Spanish bluebell *H. hispanica*, often grown in gardens, which outcompetes our native variety ruthlessly.

Buttercups

For a month in the spring we use so many perennial meadow buttercups that Fabrizio cuts them for me with a scythe. Perhaps second only to the ox-eye daisy on the definitive British wildflower list, it is tiny, bright, has no scent, and reminds everybody of childhoods spent gambolling in meadows (even if they themselves never gambolled anywhere near a meadow).

Buttercups are fragile, would never travel out of water, and drop their petals if treated roughly, but will stand a week in water as cut flowers if treated nicely – and one single buttercup in a wedding bouquet is like a drop of magic to cheer the most nervous bride.

Cow parsley

Early in the season, before the ammi and the orlaya flower, there is wild, perennial cow parsley to give you the lacy umbellifer edging that florists need – it provides so much structure and life, dancing above flower arrangements or edging tulips and narcissi deep in a bouquet. Cow parsley stands for a week in water. We don't grow it in beds, as it is prolific in the wild edges of our fields.

I'm writing this in early January, and every day as I walk my dog I see the cow parsley leaves snuggled close to the ground, waiting for an opportunity to leap. They promise spring – and great buckets of frothy white lace for weddings, for bouquets, for mixing with apple blossom and black tulips and forget-me-nots in spring bouquets. My London customers *love* bouquets with cow parsley in them, because they don't see it in the abundance that country people do. So don't weed out your cow parsley seedlings when you see them: move them to a controlled area if you will, or, if you're like me, just make the most of them wherever they pop up.

'Ravenswing' is a dark-leafed cultivar (see Chapter 4, page 75), but if you have the wild plant in your hedge they will flagrantly interbreed, and you'll find that the Ravenswing will revert to green.

Cowslips

These wild cousins of the primrose grow well as a perennial in a bed as a cut flower. Although it's not especially tall, you can cut 30cm (12") stems and they will grow in water. The flowers smell like Earl Grey tea, which with scented late-season narcissi and the sugary perfume of home-grown tulips makes for a real scent bath. Plant fresh seed in summer and you'll have good-sized plantlets to bed out in the autumn, which will flower

Cowslip seedlings grown in Rootrainers, now ready to be planted into meadow or bed.

for you the next spring. Cowslips are perennials and so will keep on flowering for several years.

Take the seedheads and shake them over your meadow or wildflower patches to encourage them to spread – they seem to do particularly well in orchards. A florist once asked me in an amazed voice how I managed to get cowslips to stand in water without wilting. Puzzled, I answered, "I cut them. And put them in water."

Forget-me-nots

These diminutive blue dots (sometimes pink dots, and very rarely white dots) of biennial and ruthlessly self-seeding flowers are fantastic when cut. They're longer-stemmed than they look, especially later in their season. At first you might think they're too short to bother cutting – but they stand brilliantly out of water and so make good dots of blue in buttonholes and children's posies. And they keep growing in length, so that towards the end of their season they might be 30cm (12") long, which is plenty long enough for me to find

Forget-me-nots are surprisingly useful cut flowers. they grow as tall as 30cm (12"), if well-conditioned they don't wilt, and a tiny sprig in a buttonhole is enchanting.

a use for them. Forget-me-nots, like buttercups and ox-eye daisies, tug people's heart strings with memories of innocent childhoods: they're certainly worth cutting. They seed themselves *everywhere*, and so we just condense the seedlings into useful clumps as we come across them through the early spring, and then cut them when they flower.

Harebells

In our clay soil we could never grow that delicate drop of heaven that is the wild harebell. If you could, in any quantity, they'd make wonderful cut flowers because (well-conditioned) they don't wilt, and the sun shines through their paper-thin petals as if they are tiny fairy field lanterns. Imagine just three harebells in a bride's bouquet. . . I've never had them to use. If you have the right conditions, then I challenge you to grow them in useful quantities and I will very likely try to buy them from you.

Knapweed

The bright purple flowers of this perennial meadow plant will develop nicely in bouquets if you cut them when they're just beginning to open. It's easy to propagate, and well worth establishing in your meadow.

Meadowsweet

Grows prolifically in hedgerows and so give it a similar habitat in your patch (slightly damp, slightly shaded) and this deliciously fresh-straw-scented, tall perennial will do well. It's useful in bud and in flower, though indoors the flowers can shed a little. I love it best for weddings.

Ox-eye daisy

Be wary of putting too many of these into water in the house, as they make the water stink as though a cat's been spraying. However, these perennial wild daisies have perhaps the definitive wildflower 'look'. They grow very well here on our damp clay, and would take over their flower beds entirely if I let them – they can bully. A daisy or three in a bride's bouquet instantly gives it the just-cut-from-the-meadow look.

Purple loosestrife

if you have boggy clay or damp ditch edges, or a damp patch somewhere in your garden, then purple loosestrife will grow well for you. It is a shortish-lived (five years) perennial, so you'll need to keep propagating a little to keep your crop refreshed – you can split plants, or it grows easily enough from seed. The tall spikes of dark purple flowers interspersed with dark green leaves are a brilliant filler with late-season dahlias or sunflowers. I wouldn't be without it.

Ragged robin

While a lovely perennial in the meadow, this is a tricky one to use in floristry, as, for me, the raggedness which is so charming in the field is too

Incredibly beautiful in the field, ragged robin doesn't transfer very well to a cut-flower bouquet – though you may have more success with it than I do.

battered-looking for cut-flower use. But it's so beautiful it might be worth a try. . .

Red campion

Cut these gently, as their brittle, hollow stems bruise easily. We have banks of perennial red campion under hedges all around Common Farm and I cut hundreds a day in season, especially when they're first in flower. They're not showy, but their subtle, velvety, dark pink flowers mix so perfectly with the other spring crops we have: a bouquet of aquilegia, buttercups, white lilac, red campion, bistort and sun spurge is as pretty a thing as I'll create in the year.

There's a lovely white variety of wild campion too (called white campion). If you grow it as well as the red, keep them apart, or they'll interbreed and you'll end up with a less intense pink and no white.

Red valerian

Valerian stands well in water, and there's a pretty white variety of this perennial wildflower (though the species name is still, confusingly, 'red valerian'), which is very useful for wedding flowers. Depending on where you are in the country, your red valerian will be a different shade: I prefer the Cotswold, slightly rusty-red-edged pink to the flat pink of our Somerset valerian, so I grow the white variety in beds and use it a great deal throughout the summer.

Rosebay willowherb

These tall, graceful, hot-pink flower spikes stand well in water and are stunning in July arrangements. This perennial will bully its way through your garden, taking over if it likes the conditions, so be wary of growing it unless you really intend to pick every flower head and only let it set seed if you want more. There's a lovely white cultivar, still thuggish if you allow it to be, which works very well for weddings.

Scabious

Wild scabious can be difficult to germinate sometimes, but once a seedling's established it will grow into a good, short-lived perennial plant. Scabious are typical of wildflowers in that they take up quite a lot of room in a bed for the number and size of flower heads you get in return. But again, think of what you owe to nature and grow them for the butterflies and the seedheads, as well as for the buttons of perfect violet colour in your cut-flower arrangements. The big scabious cultivars will give you more bang for your buck, but if you have space, the wild originals will give a subtler edge to the cut flowers you grow – and the photographs you take of your scabious aflutter with butterflies will make great posters to advertise your wares.

Snake's head fritillary

While tough to establish in a meadow, this lovely flower (pictured on page 92) is not so difficult to grow in a bed from bought-in bulbs, so long as the soil is well drained, though moist. The unfurling flower heads are caviar to a hungry mouse (annoying truth), but the flowers grow taller than you'd think (30cm/12") and the glorious chequerboard, bell-shaped flower heads are astonishing in a spring arrangement. These are certainly not a flower you'll find being imported in their millions from far afield, so grow them!

Give them some support to grow through, as they're used to growing through grass, which supports their slim stems and heavy heads. If you're growing them as a crop, a low (25cm/10") level of horizontal netting will do the trick. Watch out for lily beetle on your fritillaries – nasty red

Field scabious will be beloved of you for buttonholes – and beloved of all the invertebrates too.

beasts; at least they're big enough to see and squash (please excuse violence).

Fritillary bulbs are inexpensive, though you need a lot to make a good crop. Or you could grow them from seed, as Christopher Lloyd's mother did at Great Dixter, where her fritillaries have become one of the great draws of the garden's spring season.

Sorrel

We have perennial wild sorrel growing of its own accord in our meadow, and I admit it wouldn't have occurred to me to use it as a cut flower until I saw the sun shining through the seedheads next to buttercups and daisies, and – suffering itchy-scissor syndrome as I do – I had to cut a bit and see how it did. The subtle rust-edged green of the seedheads goes brilliantly in cut-flower arrangements, and I use it and use it for about three weeks every end of May / beginning of June. Of course, young sorrel leaves are delicious finely chopped in omelettes and with pasta, though they become very high in oxalic acid as the summer progresses, which makes them incredibly bitter, so eat them young.

Teasel

Grow this perennial for the birds – and use it green for wedding flower arrangements, and dried for your Christmas decorations. It'll grow happily along a hedge edge, though we've found it easier to establish by growing it as individual plants and planting those out, rather than expecting seed to find its way through the meadow thatch, and germinate.

Thistle

Thistles are a rich source of food for moths, and if you leave a patch to go to seed you may find charms of goldfinches feeding on them in the winter.

Wild carrot

With its gorgeous late-summer lace-cap flowers, distinctive seedheads and useful strong stems, this is a very popular flower for cutting. Grow it as a biennial, sowing seed in early June for a good crop the following year. There are pinkish varieties too, though I'm not sure that the pink isn't a little dour in colour. As always, it's a matter of taste where cut flowers are concerned, and you may find you like to grow the pink and can use it in all sorts of ways that I've never thought of.

Wildflower cutting, conditioning and use

Cut wildflowers early in the morning, directly into scrupulously clean buckets filled with fresh water. Don't overcrowd the buckets, but let the flowers settle in deep water, to have a good drink before you use them. If you're selling bunches of, say, ten stems, whether at market or to florists, strip the flowering stems of foliage before tying into bunches. Wildflowers often have more delicate stems than ordinary cut flowers, so when stripping the stems take care not to bruise or break them. It's no good trying to condition a bruised and battered stem of wildflower. Throw those away. You'll learn quickly enough how to strip foliage without hurting the stems.

Wildflowers will generally stand well in water, but they are, on the whole, more fragile than sturdy flowers cultivated for cut-flower use. In arrangements using water they should last as well as any other cut flower, though you might be wary of using too much wild material in flower-foam-based arrangements if you expect them to last more than a day or so.

Many traditional high-street florists say that wildflowers don't last when they're cut. The truth is that wildflowers don't travel well out of water, and so are inconvenient for a florist who is used to flowers being delivered flat in boxes, dry and needing reconditioning in water. A box of buttercups, for example, treated this way would arrive bruised and wilting – an unrecoverable mess.

Five top tips for growing wildflowers for cutting

❋ Grow what grows well in your area: I can't grow harebells on my clay, and so I'm not going to waste my time trying. If you're on chalk downlands, grow harebells and I will envy you from afar.

❋ Cut wildflowers first thing in the morning directly into clean water: wildflowers stand well in water as a rule, but cut them in the heat of the day and they will sulk.

❋ It takes time to establish a meadow, but many wildflowers will grow happily in a flowerbed, so do grow them as you would any other crop.

❋ Experiment: the list I've given you is by no means exhaustive. Try kingcups, bistort, wild iris – there are so many species to experiment with.

❋ Remember your tithe to nature: growing native wildflowers will feed the invertebrates and therefore the whole wild food chain in the little ecosystem that is your cut-flower patch.

Growing wildflowers increases the stock of the natural seed bank in your area. If we all grow a few wildflowers, then there will be a chain of wildflowers, food and forage for all our wildlife to live on across the land. If you want to make a living from the land, you must be kind to it and all the creatures who live upon it.

Chapter twelve

Cutting, conditioning & presenting cut flowers

So you've grown an amazing crop, but are nervous about it wilting at the sight of your scissors? This chapter should put your fears to rest – with simple steps to follow for successful home-grown floristry.

These tall, thin green buckets are called liner buckets on florists' sundries websites. They are inexpensive and invaluable.

Cutting and conditioning, for me, is the fun part of growing a cut-flower crop. You've worked and worked and worked and worked: your planting schedule has been kept up to date, you've fed your ground with manure, compost, nutrient teas, seaweed, rockdust, mycorrhizae. . . You've weeded; you've rotavated; you've lifted, split and replanted. You've kept on top of your perennials so that they don't bulk up so much that they forget to flower; you've spent a fortune on bulbs and shrubs from which your cuttings are yet to root. So, when are you going to start having some fun with this cut-flower patch?

Well, I think that the cutting-and-conditioning part of your job is the reward for all your labour. Although I seldom 'wander' anywhere, I feel that 'wandering' is still part of being a flower farmer. At 5 a.m. on a still June day, when my dog Chocsy and I set off across the meadow with bucket-

packed trolley rattling along behind us, scissors in pocket and list in hand, and the barn owl's out hunting with his *whumph* of wings, and the flowers have no scent yet in the early-morning cool – then the world is mine and mine alone. I feel I've won something precious.

As Chocsy thunders across the meadow in her not-very-lightfoot way and my trolley begins to fill with the flowers I need for the wedding, or bouquets going out by courier that day, or for my local shop orders, I am supremely grateful, to whatever combination of magnetic field + chemical reaction that drives us, that I have been driven outdoors at that very moment. To have the world to yourself in the limpid dawn light, to stand in that world between rows of thousands of sweet peas and watch the dew shake off their ruffled petals as I cut them – that's why I do this job. To be forced to get up early for any other work would be no joy, but to be forced to stand in a meadow – the sun still cool, the dew still wet, the world still quiet – is heaven.

And to be able to exercise your creativity with what you have grown. . . I won't retire a millionaire, but with a job like this, will I ever want to really retire? (She says, rubbing sore feet with balm. . .)

So, what do you need to make sure that all this early-morning communing with nature goes smoothly and, at the end of the experience, you have what you need for the day?

The flower-cutting toolkit

Florists often use a real toolbox for their kit – a good idea which I plan to emulate one day. We keep our scissors hanging in orderly rows in the flower studio; the raffia hangs from a hook in the ceiling; the cellophane is on a roll for easy dispensing; scrubbed buckets are indoors, buckets to be scrubbed out are outdoors. . .

Clean scissors

I use florist's carbon-bladed scissors (of which we have about 20 pairs hanging about the place at any one time). Use secateurs to cut a great many flowers and you'll be suffering from repetitive strain injury within a week. You only need your secateurs when cutting thick, woody material. Cutting flower stems is a snip-snip-snip operation. You'll be cutting hundreds at a time of each variety on your list. Believe me: florist's scissors over secateurs will save your hands from arthritis and your arm from tennis elbow.

Your scissors need to be clean to prevent the spread of disease: don't move from cutting a lot of mildewed ranunculus with one pair of scissors to cutting sweet peas with the same. Keep disinfect-

How to cut flowers

❋ Banish any fantasy you may have had of wandering about your garden with a trug and a pair of secateurs enjoying the sunshine. Flowers won't thank you for being cut in the middle of the day when the sun is high on your (and their) head. Cut them early in the morning or late in the evening, when you can't feel the warmth of the sun on your neck.

❋ Always cut flowers direct into water: you may have a vision of yourself elegantly swanning about the garden in a picture hat cutting flowers into a trug, but if you take this approach your flowers might struggle to rehydrate when they get them back into the house. So, take buckets of water out with you to cut your flowers straight into. Cut the flower, strip the foliage, and plunge the stem into a bucket of clean water before it's had time to dry out.

❋ Flowers 'drink' water through the capillary action provided by the spongy cellulose in their stems. Don't give that cellulose sponge time to dry out before getting the stems in water.

❋ Cut woody stems at an angle, and also snip them up the stem as you cut the stem from the plant, to increase the area from which the flower can absorb water (see illustration in Chapter 7, page 124).

ant to hand – either thin bleach in water or a spot of white vinegar, which works just as well – and wipe your scissors down often.

Keep your scissors sharp: sharp blades cut through plant matter cleanly, so helping the plant recover quickly from the wound. Blunt scissors squish and bruise and leave untidy ends, not only on the plant but also on the cut flower, which is now conditioning in a bucket of water – where bacteria will breed more easily if the cut ends are smashed rather than cleanly cut.

Secateurs

You will need secateurs: for cutting woody branches and so on. Don't struggle to cut thicker stems (do as I say, not as I do) with florist's scissors, as you'll only strain your hands, blunt your blades and make a mess of the stem you're cutting.

Clean buckets

So many buckets to choose from, and only one stricture for them all: they must be clean if you're going to condition cut flowers in them. And I don't mean just rinsed out by the greenhouse tap – I mean scrubbed with washing-up liquid and a spot of bleach.

Bacteria is the enemy of long-lasting cut flowers. If you stand flowers in fresh, clean water, in scrupulously clean buckets, then they will condition beautifully. Stand them in clean water in dirty buckets and the bacteria in the buckets will rejoice in the arrival of fresh-cut green matter, and your flower stems will not last.

Bucket sizes

You must see which suits you. Ordinary black yard buckets – the sort you buy at garden centres or saddleries – are too shallow for their width for conditioning flowers. Very long stems tend to be heavy-headed and can pull themselves out of these buckets.

Sometimes supermarkets have cheap buckets on offer: they are useful (and very cheap), but often

so flimsy that they can't be picked up one-handed when full of water without cracking and spilling flowers and water all over the place.

Florists' sundries websites list any number of size and width of buckets. I tend to use more of the tall, narrow, green 'liner' buckets, as they're called: while they take a useful number of flower stems, their narrowness means that they don't take up a lot of room (when cutting for a wedding this is a useful attribute for a bucket), so that I don't end up putting two or three different varieties

Cheap tin buckets and jugs are attractive, but often leak. I hide a cut-down plastic water bottle inside these jugs in case of accidents.

A trolley, to fill with clean buckets of fresh water into which to cut your flowers, is one of your most useful tools.

of flower into one bucket, and their height means that heavy-headed flowers are kept upright, so I don't risk buckets falling over.

You could use a basic mixture of bucket sizes. So, for a first season with three raised beds and a modest output planned, you could buy yourself three tall, heavier black buckets with handles, ten tall narrow green buckets, and also a few shorter containers for your short-stemmed flowers.

Beware tin buckets from florists' wholesalers: they can be leaky if you over-fill them. Their weak points are at the joins, where there's usually a ring about three-quarters up the height of the bucket. Don't over-fill them, especially if you're using them for display. When they arrive, test them for leakage before using, especially if you intend to sell flowers in them.

Trolley

A trolley for pulling your buckets along to your cut-flower patch is indispensable. With a trolley you can count in advance how many stems you're going to cut, know how many buckets you'll need, have them filled with water, and be cut, back at the house and ready for breakfast in no time. Without one, you'll be constantly lugging buckets back and forth. We didn't have a trolley in our first year (I thought it too expensive), and I remember the days of slogging back and forth with a bucket of flowers in each hand with some amazement that I would put myself through that.

The list

I make a daily list of what needs to be cut. I count the stems in advance: 20 of this, 50 of that. I know how many bouquets I'm cutting for, and I have a fair idea of how many flowers I'm going to need through the week. We cut each individual bouquet to order, so that flowers are never languishing in a barn waiting to be used.

Cut the number of stems you have on the list you've taken with you: this way you'll avoid wastage.

Ensuring your flowers are in best condition

Conditioning flowers is what you do to get your flowers in the best state for floristry. The time they spend in buckets of clean water, immediately after being cut, is the time when they are being conditioned. They should have a minimum of four hours' conditioning if they're cut straight into water and will be put back into water (for example, aquapacked in a bouquet) when they've been arranged. If flowers have spent any time out of water (as with the flowers we have sent up from Cornwall when we need extra stock, which travel overnight out of water, flat-packed in boxes), they need a good 12 hours to rehydrate before being used in floristry. Flowers that have been out of water for any time need their stems re-snipping (to open the spongy drinking cellulose cells again) before they're conditioned.

If you're not using flower-preserving chemicals (see opposite), then you need to cut to order and as close as possible to the date that the flowers will be needed. Remember not to crowd your buckets with flowers, so that they have air as well as water while they condition. If you cut in the evening, then a night in water in a cool, dark, airy place will give them plenty of time to have a good drink, ensuring they're in top condition when you arrange them the next day. If you cut in the morning, they can have all day and night conditioning.

Cut flowers with as long a stem as possible. Whether you're selling them on in bunches of ten to a florist or planning to use them for your own floristry, long stems give you choice, and will make the stems individually more valuable.

If woody stems are cut shorter during the floristry process after conditioning, then snip them up the stem again to give them maximum surface area for drinking through.

Have somewhere cool to keep your flowers – whether they're being conditioned or have been arranged for an event or fair the following day.

Flowers will need time conditioning to recover from the cutting process. Here there is enough material for five bouquets.

Searing

Woody stems and a few field flowers condition better after a little searing. Take a tin bucket out to the shrub you plan to cut from, with 5cm (2") of boiling water in it. Cut the stems from the plant at a clean angle. Cut up the stem again 2-3cm (1"). Plunge the stem into the boiling water.

In the time it takes you to cut, say, 20 stems into the boiling water, the stems will have had time to be seared. Then fill up the bucket with cold water and allow the cut stems to condition as normal.

What to sear?

Any woody stems you want to cut from shrubby foliage may do better for you if you sear them.

Poppies will stand in water with a little searing. But remember, while searing poppies will enable them to last in water for a day or so, a poppy won't ever last longer in a vase than it does in the field, which is fleeting.

You can also cut stems of any flowers that have wilted during the conditioning process: they may revive with a freshly snipped stem and 30 seconds' searing in boiling water.

I will admit I only really sear poppies and lilac. If you cut your material early in the morning or after the heat has gone from the day, then most flowers and foliage will happily cut straight into cold water and condition overnight.

A word on flower-preserving chemicals

If you are growing flowers in a small plot, on a small scale, for a local market, then I would

Wedding flowers should not be inexpensive. They should fairly reflect the cost of growing, cutting, conditioning and arranging them.

question the need for chemical flower-preserving liquids or powders. Indeed, I always question the need for a sachet of anything where the ingredients aren't listed – and while I can guess that most commercially available 'flower food' consists of a mix of bleach, sugar and some pH-balancing agent for the water, unless I *know* this I won't use it. We pour the water from our flower-conditioning buckets straight back into the ecosystem, and I have no fear that it'll do any damage, because I've added nothing to it. Of course, what you treat your flowers with is your choice, but I urge you to think before using flower food or chemicals just because somebody's told you you must. Flowers from your plot will be cut direct into water and delivered to the customer four or five days fresher than flowers flown in from abroad. You can guarantee that yours will last a week without the aid of flower food.

If you would like to make your own 'flower food', however, you can add 1 teaspoonful of sterilizing powder, the juice of half a lemon and a teaspoonful of sugar to 1 litre (1¾ pints) of water. See Chapter 9, page 152, for a recipe for sugar solution for feeding sweet peas.

Flowers that spend a lot of time travelling between farm and end user are often treated with a solution of silver nitrate, which stops them producing ethylene, the same gas that bananas and other fruit produce when ripening. If they are stopped from producing ethylene they'll last a few days longer in the vase. You know when you buy a bunch of lilies from the supermarket, take them home and put them in a vase, and they never open, and never open – and eventually, about two weeks after you bought them, the ends of the still-unopened buds turn brown and you throw them away? Those lilies have been over-treated with silver nitrate, and it's stopped them developing at all. Without this treatment, flowers won't last so long – but your flowers will be reaching the end user 4 or 5 days fresher than those flown in from distant lands, so you can leave them untreated. And the very fact that they've not been pulsed with silver nitrate can be a sales point for you, because your flowers will grow in the vase, releasing their lovely scent (in the ethylene they produce) throughout your customer's house.

Cutting for weddings and other events

If you're cutting for an important event such as a wedding, make up a sample of your table centres (for example) in advance, and count the stems you use in it. You can then multiply that number by the number of table centres you will need, and cut accordingly.

Beware of cutting 'some' of each kind of flower you'd like to use for the wedding until you have 'enough'. With a list of numbers, you won't a) find that you're drastically short of material and therefore have a panic at the last minute, or b) find you've wasted time massively over-cutting, resulting in loads of material left over at the end with no destination other than the compost heap.

No matter how small the event, plan to cut your flowers with no wastage. When a new customer comes and asks you to cost up a wedding, refer back to an exact list of how much material you used for a previous event to help you work out exactly what you'll need.

People often think that home-grown flowers should be inexpensive. Remember, you've bought or rented the land, nurtured the plants from seed, been up at 5 a.m. to cut the flowers . . . all this before you've even started arranging them, if you're going to do the floristry too – and the chances are that for an event you'll be starting at dawn on a Saturday and may well be paying someone else to get up early and work with you. No, wedding flowers should not be inexpensive. They should fairly reflect the cost of growing, cutting, conditioning and arranging them.

Vases and containers

For me, this is another fun part of this job. Keep an eye out at charity shops and brocantes for good vases, but also for jugs and coffee pots, teapots

Floristry is as much about presenting the bouquet as about making a lovely posy. A fine finish shows that you care and adds value to what you do.

and any other interesting containers. But remember to buy vases too! Not everybody wants vintage teapots for their flowers. You might also find that you can hire out good vases for weddings, with your flowers in them.

If you plan to sell your wares at a farmers' market, they'll look cheery displayed in interesting vintage crockery. Farmers' market customers tend not to want to spend much: they like to buy ten stems of larkspur, for example, and twenty of sweet peas, rather than made-up posies. But if you do make up some posies and put them in an attractive container, then you'll draw people to your stall, show them what they can do with the flowers they buy from you, and also give people ideas for wedding and event flowers.

Collect attractive glass and china for displaying your flowers. You can also hire them out for events with your blooms.

Wire and supports

We keep a good selection of different-gauge wires in our floristry supplies box. We use reels of thin wire for garlanding and Christmas. Thicker, short stems of stub wire are great for heavier items or for fruit that need support in arrangements. Very fine stub wire is best for wiring buttonholes and corsages. Keep a pair of pliers in your toolbox as well, for pulling wire through when it's too hard to get hold of with fingers.

If you're arranging wedding or event flowers, then a collection of narrow sticks will be useful to prop up hollow-stemmed flowers that need to remain straight and tall in big arrangements. From wooden barbecue skewers to the thinnest-grade bamboo, you may find that sticks save your floristry from time to time. Of course in winter you can use budding sticks cut straight from the tree to provide a support system through which you can thread your flower arrangement. The skill with floristry is often to think outside the box in order to turn the vision you have for your flower arrangements into reality.

However, I always remember my mother's advice: "Never try and make a flower do what it doesn't want. Because it won't, and you'll have wasted half an hour losing a battle." Wise words.

I will admit we don't use a great deal of sticks up hollow centres with our floristry. I like my flowers to curl into interesting shapes and convolute towards the sun if they want to. Our floristry is not the dead-straight sort. You, though, may dream of an honour guard of delphiniums along window sills for church wedding flowers. Equip yourself to meet your needs.

Flower foam

Flower foam is another contentious issue where floristry is concerned – competing hard to be one of the least biodegradable products of this oil-

Make a list of florists' sundries that you'll need when you start out: buckets, scissors, wire, ribbon, pins. . .

I always remember my mother's advice: "Never try and make a flower do what it doesn't want. Because it won't, and you'll have wasted half an hour losing a battle."

based, carbon-hungry age. Florists sometimes look at me in amazement when I explain that it will degrade no faster than your average plastic bag and then say, "Oh, so that's why it always comes out of the compost heap in one piece!"

So don't try to compost your spent flower foam – it won't biodegrade in the slightest. Think about whether you need to use it at all. We do use flower foam in our floristry, but if there's an alternative, we'll suggest that to our client first. You could make it part of your USP (unique selling point) that you don't ever use floral foam. You will find this quite a draw for some customers.

Practise using wire and tape rather than flower foam, and make a collection of flower frogs. A flower frog is any kind of tool that means you don't need to use flower foam: this might be a circle of stand-up pins on a heavy metal base, a container-within-a-container into which you can secure stems, a simple ball of chicken wire, or a layer of wire over the top of a larger container into which stems can be secured. We often avoid using flower foam by making grids of tape on top of a large container and then use a heavy pin wheel at the bottom of the container.

When is flower-foam use inevitable?

Well, for big garlands for weddings and events we find that foam means we can use any kind of cut flower without fear of wilting. You can't always arrange the flowers at the very last minute, or at least you can't always arrange *all* the flowers at the very last minute, and garlands use a lot of material. For us it's easier to use a flower-foam-based garland on these occasions, although the rest of the flowers for the event may be in vases and not need any foam. So, don't assume you're going to need floral foam: as a rule of thumb, think first about how you would do flowers for an event without it.

Alternatives to flower foam

Moss is one alternative to foam. It's illegal in the UK to forage for it unless you own the land it grows on, but the moss from your own garden is yours to use as you will. You can buy moss from Cornish suppliers. Fill a chicken-wire frame with moss, soak it, and use the result as though it were a flower-foam-filled frame. You could also try soaked shredded newspaper framed in chicken wire. You might need a sharp metal kitchen skewer to make holes through the newspaper in which to put the flower stems. If you're in a bracken area, then soaked bracken works as well as moss for stuffing frames in this way.

Hanging baskets can also be used as a framework: stuff two hanging-basket frames with moss (or newspaper or bracken) and wire them together to make a 'moss ball' for wedding or event floristry.

If you're using moss to act as a water reservoir in floristry, then you might need German pins: U-shaped pins which secure flowers and foliage in place.

It's worth remembering that flowers last a great deal longer in water than in flower foam. This is a potential sales point when talking to a client who might assume that flower-foam-based arrangements are the only option.

Keep a box of small test-tube-shaped containers (available from your florists' wholesaler) in your floristry cupboard. These are useful for posies on pew ends and chair backs that need a little water. Equally, tiny posies in test tubes hung in trees at weddings are very pretty.

Binding

Here are some useful ideas for binding. As usual, my list is by no means exhaustive, but offers a good starting point for you:

* Raffia: good for binding buttonholes as well as tying bunches of flowers. Elastic bands take less time to use, but look as though they take less time and are not very eco-friendly. Think about your business image and maybe take the time to use raffia or garden twine rather than elastic bands for tying.
* Garden twine: an alternative to raffia for a buttonhole or posy. There's a lovely company called Twool that makes garden twine from British sheep's wool.
* Stemtex: this is tape that feels a little like crêpe paper but which is coated with a tiny film of glue, making it stick to itself when you bind buttonholes or bouquets with it. It's useful to an extent, though if you use other materials such as raffia and string to bind buttonholes, you may find that one box of Stemtex lasts an awfully long time.
* Ribbon: a collection of different-coloured ribbons is useful for finishing bouquets (and fun to source – another treat).

Finishing

The 'finish' you apply to your bouquets or bunches will frame them for sale. It's really important to consider what and how much finish you will supply, as every time you twist raffia around stems or wrap paper around a bouquet you're adding to your costs and to the price the customer pays. But consider too whether the customer might prefer to pay a little more for beautifully presented flowers. Might it be the finish that makes the sale? Your basic materials for finishing should include:

* Tissue paper: a collar of tissue paper finishes a bouquet beautifully.
* Cellophane: if you want to make 'aquapacks' for your bouquets (essentially bags of water in which your flowers can be kept fresh), you'll need cellophane.
* Pearl-headed pins: these are to floristry what safety pins are to the household toolkit. Useful on so many occasions, often when you least expect it. You'll get through one box of pins surprisingly quickly.

An alternative to cellophane, which is available on the market, is a sort of bag lined with disposable

At the end of a busy day your feet will ache and you'll have earned that cup of tea.

nappy material, which will absorb water to keep flowers fresh (not very stylish but practical – as ever, it depends on your market and how eco you want to be). Gel pearls are an option for keeping flowers hydrated, and come in a pleasing array of colours. Personally I wouldn't touch them with a bargepole, as they seem to me to be the very antithesis of our natural, fresh-from-the-garden look and ethos, but sometimes you may feel that practicality has to be taken into account and compromises have to be made.

Five top tips for cutting and conditioning flowers

❋ Cut flowers at a time of day when you can't feel the sun on the back of your neck.

❋ Use scissors rather than secateurs to cut flowers, for a cleaner, sharper, kinder-on-your-hands cut.

❋ Keep your buckets scrupulously clean: breeding bacteria in water is the enemy of the cut flower.

❋ Cut flowers directly into buckets of fresh water: never let plant material lie about out of water and expect it to do well in floristry afterwards.

❋ For best results, let cut flower stems condition overnight before using in floristry.

Chapter thirteen
Hedgerow Christmas

If you think artisan flower growing is a job for just the summer months, then think again! You'll be surprised what you might find to use in floristry in your deep midwinter garden, especially around Christmas. Deck your halls, and those of your customers, with the fruits of your hedgerows and meadows.

Dressed wreaths are very popular at Christmas markets.

However greenery-based and seasonal they may look, shop-bought Christmas decorations are largely imported. Here in the UK we get Christmas trees from Scandinavia, forced bulbs from Holland, and roses (we love roses all year round) from South America and Africa.

But you can easily keep your Christmas entirely locally grown. Using hedgerow greenery you can supply your local farmers' market, farm shop, home interiors shop and florist, not to mention your own customer base, with gorgeous home-made decorations for Christmas. Even if you set out to be a summer-season-only grower of cut flowers, if you build up a nice customer base over the summer months, I'll bet you they ask what you're planning for Christmas. So have some ideas and be ready for them.

If you've had a stall at a farmers' market over the summer, your customers will quite understand if they don't see you for a few weeks between, say, the end of October and the beginning of December. But be sure not to miss the run-up to Christmas: there will be Christmas specials at your farmers' market and bespoke Christmas markets springing up all around your area, where you can not only sell wreaths, pots of bulbs and so on, but also begin to drum up a bit of business for the following summer.

Disappear for too long and your customers will forget about you. Pop up when they're out shopping for Christmas and you'll remind them that you're there. And, at this key time of year for any retailer, you'll come across lots of other customers who might be interested in what you're doing next year. Farmers' markets are not places to make a fortune, but they are brilliant marketing opportunities for anyone without a shop. So, make Christmas wreaths and market them!

Ingredients for a home-grown Christmas

As the summer passes, keep an eye on your garden for useful material to dry for wreaths and garlanding. Winter is a time when people often have a big clear-out in their garden, so keep an ear to the ground for people clearing ivy and cutting down holly – a great resource for the florist.

And remember: no two Christmases are the same. One year the spindle berries will be so abundant that your wreaths will all be pink and orange; the next you might be given the top off a variegated holly tree that someone's cutting back; another year you may find you have an incredible supply of mistletoe. Be ready to ring the changes in your Christmas look. Pre-sell wreaths as 'No two ever the same'! You'll need to be light on your feet and think laterally and creatively to make the most of the material at hand.

And don't forget to think about potting up bulbs for your Christmas customers (see Chapter 5).

A willow wreath dressed with a garland of mistletoe, ivy, spindle berries and other lovely Christmas ingredients.

Festive floristry

Your list of ingredients for Christmas decorations might look something like this:

* Pine, holly, ivy
* Euonymus (decorative, variegated)
* Old man's beard
* Rose hips
* Any red berries you have in your garden
* Crab apples
* Mistletoe
* Spindle (also a euonymus, but here I'm speaking of the gorgeous pink-and-orange berries, which make such wonderful wreath ingredients).
* Hydrangea heads: can be hung to dry from any time in the summer.
* Statice: grow it in the summer and dry it for Christmas wreaths – it's a brilliant wreath addition. I particularly like the white, but you might like splashes of pink or purple in your wreaths.
* Honesty seedheads
* Poppy seedheads
* Nigella seedheads
* Wild carrot seedheads
* Scabious seedheads
* Teasel
* Rosemary
* Box and Christmas box
* Willow (see page 201 for more on this)
* Clematis prunings: they weave nicely to make a base for wreaths and garlands.
* Any other easily obtainable decorative bits and pieces, such as pheasant feathers.

Preparation

As with most things, the more prepared you are, the more you can achieve. Think backwards from Christmas. How much money do you want – or need – to make to cover your costs? Think carefully about the cost of your wreaths: there is always a balance to be struck between your desire to sell lots of your product and your need to make some money from it

Small, inexpensive wired wreaths can be bought on the high street for as little as £8 or £12. Are you in this market? Are you just going to make simple wreaths from your own greenery, for sale cheaply at farmers' markets? Perhaps you'll offer heaps of greenery for sale, for people to take an armful home with them – mixed bunches of holly, ivy and clematis clippings to garland with? Or do you want to indulge your creativity a little further? Larger, dressed wreaths are a more difficult sell,

but you might get more satisfaction from making them, allowing your artistic juices to flow. A stall hanging with lots of lovely different wreaths is a great draw at a market or a Christmas fair.

A few 'special effects' go a surprisingly long way in a Christmas wreath. A wreath made with ten whole hydrangea heads is wonderful (if a little 1950s-swimming-hats-in-a-circle-ish). But a wreath into which small flourishes of hydrangea flowers, cut into a lighter promise of next summer, are mixed with variegated myrtle and explosions of orange 'Rambling Rector' rose hips, is more delicate, subtle, and much more interesting to look at.

Hedgerow Christmas toolkit

Your hedgerow Christmas toolkit is a variation on your everyday florist's kit. In early November it's worth checking that you've got plenty of everything you'll need, as you don't want to run out of wire halfway through your Christmas orders and have to stop production for two or three days while you wait for delivery of a new batch.

- Florists' wire: several kinds, although we mostly use reels of thin wire – as it's on a reel, we don't lose the ends of our garlands as we're working.
- Pliers: for pulling recalcitrant wire, but also for pulling willow and other strong, bending bits of natural material through spaces where fingers numbed by work and winter will not perform.
- Secateurs: there are always woodier stems at Christmas than those your florist's scissors will easily handle.
- Ribbon: not just red ribbon, or green. Think: orange goes beautifully with a lot of berries; gingham is lighter than heavy double satin; raffia makes strong bows and comes in more than the natural attractive straw colour.

Match your ribbon colour to that of the berries in your garland – as here, with orange 'Rambling Rector' rose hips.

* Christmas is a time for decoration and colour, so it's worth building up a collection of ornamental extras – beads, feathers and so on – to add a finishing touch to your wreaths and garlands.

The willow wreath

Here in Somerset we make our wreaths out of the six or seven different-coloured willows we have growing around the edge of our smallholding. We planted them when we arrived here by striking 1m (3') lengths of fresh willow clippings harvested from a neighbour, who has a lovely willow collection which he cuts back every year. At that stage we had no idea that we'd eventually start farming flowers for a living, but our land was boggy and we thought that an edge of willow would give us a fast-growing windbreak and might drain the ground a little.

The willow rooted – and our willow boundary is now perhaps the best thing we've done here. Two or three Christmases after we moved in, Fabrizio made me a wreath out of our flourishing willow collection, and the colours of the bound willow together were enormously pleasing. I dressed it with a half-circle of hedgerow garland and hung the wreath on our front door. And so began our hedgerow Christmas tradition.

Our willow wreaths are made with only willow – no wire. The tension in the circle is provided by the willow: it holds itself in place, bound with willow lashing. And over the following year, after the wreaths are made, the willow circles dry out slowly, hanging against stone walls or on a painted barn door. Once the hedgerow Christmas garland has been removed they are no longer Christmas decorations: simply circles of willow magic adorning the house and land.

Cut your willow in advance and count the stems, so you know how many wreaths you can make.

How to make a willow circle

To make a 50cm (20") willow circle you will need:

* Nine 3m (10') lengths of willow – any different colours will do. The wild willow called 'crack willow' will be the most difficult to work. You could also use dogwood stems.
* Three 1m (3') lengths of thin willow whips for binding.

Work all the willow to soften it by pushing it through your hands, using your thumbs as softening pads. You'll feel the fibres of the willow begin to separate inside the stem as you work it. This will make it much easier to create a circle. If you don't work the willow before you use it, you risk the stems cracking as you weave them in and out of your circle. A cracked stem in a willow circle will not always show, but will weaken the tension of the wreath. With a cracked stem you won't be able to get a smooth round shape.

Once your willow's been softened by massaging it in this way, make your first circle. Your first length of willow can be twisted round by weaving the end in and out of itself. A 3m length will go round your circle two or three times. Bend the willow through the circle by putting the thin end in first and pulling the rest of the willow through, as if you were doing a giant's sewing. This will also help prevent the willow from cracking.

Your first circle will be off-centre, pulling itself into a slightly corner-ish egg shape. Don't worry, as the further eight lengths of willow that you weave into the circle will create the tension to pull the circle into a better round. As you weave each length of willow into the circle it will become firmer, stronger, and less likely to ping out of shape.

For your last length of willow, you might save a stem with lots of side shoots. This will give your willow circle a lovely explosive look, with the shoots bursting out of the willow so that it looks like a sun, or a royal race-goer's fascinator.

You can then bind your circle in three places with the thin, shorter whips. This binding will firm up any loose-feeling places. Push the woodier end of the whip through a gap in the circle's willow and hold it firm. Pull the whip round and round the bundle of willow, pulling as tight as you can. At the end, thread the fine end of the whip through the back of the binding, as if you were sewing, and pull it tight. This is where you might need your pliers to pull the whip through.

The garland

A garland is used to adorn your willow wreath, but it also works just as well without willow to attach it to. Make a sturdier garland than that described as follows, and it will hold itself into a circle to make a wreath if you'd like it to. Equally, lengths of garland do well along mantelpieces (but do be careful with open fires and lit candles nearby – garlands look great, but will dry out, and if you're using any pine in your ingredients remember that pine is super-flammable).

There are no rules regarding what you should use to dress your wreaths with.

How to garland your willow circle

You'll need a selection of hedgerow goodies and a reel of light florist's wire. Florist's wire comes in all sorts of gauges. You may find this garlanding easier with thin wire, which will pull tight like string, rather than thicker wire, which is good for handling heavy items such as fruit or heavy-headed roses but is harder to hide in the foliage and would mean over-engineering your wreath.

Cut your hedgerow ingredients into quite long lengths – 20-30cm (8-12"). You can trim them later. So, your list of ingredients could look like this:

* three lengths of flowering ivy
* three handfuls of berries
* three spurts of variegated euonymus
* a dried hydrangea head
* a few tufts of old man's beard
* three heads of dried statice
* a few silvery coins of honesty seedheads.

You want each length to be long enough for you to be able to wire each next ingredient of your garland on to the stem of the previous, so that your garland is a long ribbon of material, not a bunch. This can be difficult if your habit is to make hand-tied posies: it feels wrong to be attaching your material along the stem of the previous piece that's been used. Practise a little, to get into the swing of it. Very bunchy garlands a) use a lot of valuable material, and b) look almost as heavy as traditional weighty wreaths, and what we're trying to create here is a feeling of lightness and freshness.

Wire your garland ingredients together, one after another, down the long length you make with their stems, until you have a garland perhaps half or three quarters the circumference of the willow circle you're going to attach it to. Or perhaps you'd like to cover the whole willow circle with garland, in which case it needs to be as long as the full circumference of the circle. Attach your garland firmly to the circle with a little wire. You might find it convenient to use the bound sections of willow to thread your wire through (another reason to use fine wire: it's easier to thread through tightly bound handfuls of willow), though if your willow circle has lots of explosive hairiness you should have plenty of little places to which to attach your garland. Add ribbon to taste, and there you have it.

Five top tips for a successful hedgerow Christmas

* Remember to harvest seedheads and material throughout the autumn, which you can put by for use at Christmas.

* Think carefully about how much time and effort will go into your wreaths: just because somebody else is selling wreaths very cheaply doesn't mean you have to. Don't end up out of pocket!

* Look! Your garden is full of interesting foliage and berries, which you will see if you look for them. Ivy berries are beautiful in garlands, and you may have a whole tree of them.

* Wreathing is very hard on cold, winter-bitten hands. Use strong hand cream to treat them with at the end of every long working day.

* Remember to tell your summer customers that you'll be taking part in local Christmas fairs and farmers' markets, so that they can spread the word about what you have to sell as well as come along and buy from you.

Chapter fourteen

Starting a cut-flower business

Growing cut flowers for money is a great idea for a kitchen-table business – but even the smallest venture will work better with strict accounts kept, business plans made, and a clear strategy for its future development.

Be as exacting with your accounting, no matter how small your business, as you are with putting up your polytunnel, and all will be well.

At the end of our first year we had worked flat-out and I was stunned to find, on day one of year two, that we were £200 in the red. I had assumed that hard work and lots of sales must mean we'd make a profit. This sad fact taught me to care about my business admin.

If I can do it, you can. My background is in fashion, the humanities, writing. I have never in my life been good with money. It slips through my fingers for many reasons, but mainly because I'm over-confident about mental arithmetic and I have a bad habit of assuming the best rather than calculating the actual figures. My mathematics has improved beyond my wildest imaginings since I began the business of growing cut flowers for a living!

Running Common Farm Flowers has brought together all the skills I've picked up through my life and has made them useful. However pointless I might have thought them at the time, I am surprised again and again by finding myself doing something I learned at 21, at 30, at 40. . . Here are just a few examples:

- Presentation is all: a fact I learned at *American Vogue* in Paris, where we used to stuff the sleeves of the couture gowns we sent to be photographed in New York with tissue paper to keep them perfect. The flowers people buy from us are nearly always a gift – and presentation frames the gift, so our flowers are always beautifully packaged. People don't buy because we're artisan or organic or local, they buy because they like our flowers, and our packaging has to present the flowers to the high standard of the bouquets we make.
- A cashbook will tell you how much money you *really* have, as opposed to what you *appear* to have when you look at your bank balance. This was taught me by John Galliano's backer's right-hand man: a kind, moustachioed gentleman, who taught me how to set up systems within which Galliano's business could run – systems we now copy in a miniaturized but exact way here at Common Farm Flowers.
- Marketing is everything. When I was at *Vogue*, we girls in the editorial department envied our friends in marketing their higher wages, but didn't envy them their jobs. Selling advertising space seemed to us a thankless and creatively flat job. I see now why those girls were paid more: without them we wouldn't have had a job at all. Marketing is everything in any business, including one grown from the cut-flower patch outside your back door.

This chapter is, inevitably, somewhat UK-centric. But in any country there will no doubt be the equivalent of the British Local Planning Authority, which may have an opinion about whether you're allowed to put up a polytunnel on your land; and a government department to which you'll have to report. So, wherever you are in the world, do read this chapter, and make a list of which local authorities you'll need to check in with or ask questions of before you get going – and as your business grows.

Realistic expectations

Before you start thinking about setting up a business in cut flowers, ask yourself this question: Do you really want to grow flowers for sale? Why? For the money? Because you love flowers? Or because you love gardening?

If you think flower farming is a great idea because you can just see yourself outside, with a radio for company, spending your day digging – then think again: that is the picture you can associate with being a *gardener*. Farming your garden not only involves growing but also *selling*. If you don't fancy the sales aspect of this job, then you may be looking at the wrong career.

Gardening also makes a great career: one for which you can charge a respectable hourly rate, through which you can exercise your creativity in other people's gardens, and in which you can improve your knowledge and therefore your financial value while you work. If you fancy flower farming because it is the gardening part that appeals – and if you're going to have to rent/buy land in order to grow cut flowers for money – then gardening might (might! I don't want to put you off, but it's worth thinking about) be another option to consider.

If, on the other hand, you fancy flower farming because you love the sight of masses of flowers blooming in a garden, then, again, this may not be the job for you. The person growing cut flowers for money is a person with serious itchy-scissor

In our first year we grew this blinding mix of flowers and they sold like hot cakes, helping to build our reputation (though we're a bit more circumspect with our colour palette these days!).

At the end of our first year we had worked flat-out and I was stunned to find, on day one of year two, that we were £200 in the red. I had assumed that hard work and lots of sales must mean we'd make a profit. This sad fact taught me to care about my business admin.

syndrome, who can barely wait for a bud to develop before cutting a stem – a person who is really very uninterested in what a garden looks like, and cannot wait to get each flowering stem indoors, well conditioned and used in a bouquet.

In the four years since we started Common Farm Flowers, the learning curve has been steep and surprising. I might summarize the pros and cons of the job as follows:

Pros:
* We work from home.
* I'm going to garden anyway – so make the garden pay.
* We work creatively.
* We're self-employed: not at anyone else's beck and call.
* No commute.
* We may be working but we see lots of our kids.
* Fresh air and exercise are included in the job description.

Cons:
* Seventy-hour week.
* Every day an early start.
* No work = no pay.
* We have to ratchet up motivation daily.
* No escape from the job.
* The job invades home.
* We work outside whatever the weather.

When I teach flower farming, at the beginning of the day I ask my class how many of them picture themselves in the garden with a cup of tea, a garden fork and a radio. They nearly all raise their hands.

If you don't mind whether or not you sell from the barrow outside your gate, then by all means spend your summers gardening with your radio and your tea. But unless you have amazing passing trade, then you'll still have to fit your marketing plan into your schedule, or nobody will know that there are sweet peas for sale in the barrow outside your front gate.

Farming your garden is extremely hard work, and very much a lifestyle choice as much as a business choice. You will certainly earn more for less effort in almost any other walk of life. However, the person who farms their garden is outside, deeply involved in their relationship with the soil, the sun, the wind and the rain. . . So, I'm not for a second suggesting that you shouldn't start a business farming your garden: I'm just saying that you should go into it with your eyes wide open.

First things first

I am a great fan of a list. I'm afraid this chapter is rather a list of lists to make, but, no matter how small your operation, it's worth taking the time at the outset to think about the following key aspects of running a business.

Make a business plan

I have a friend who began what has become a large and very successful publishing company, and his one piece of advice to us when we started was: "Fail to plan and you plan to fail."

Whether you intend your cut-flower business to supply your neighbours from buckets in a barrow by your front door, or you have your eye on a bigger market, organizing your operation as a proper business from day one will not only set you up with good habits but also keep you informed as to the true cost and return of the work you put in to farming your garden. You may be planning to give all your proceeds to the village playground fund, but it's still worth knowing that from a packet of seed costing £2 you made £100 to donate.

Making a business plan requires you to be sensible, to tame your wild ideas into practical theory. It will give you a structure to work with. It's not so much about the actual amount of money you'll turn over – which inevitably involves a bit of guesswork at first – as a plan for what you'll have to do in order to achieve money turning over at all. A business plan will inform the decisions you make, highlight potential problems in advance, and make you think seriously about your floral dream as a practical day-to-day reality.

A simple business plan should include the following:

* A one-page mission statement: what do you plan to do? Where, how and why?
* A page of research facts: what have you found out about the potential market for your flowers? Where do you plan to sell? Will there be competition? How will you price your product?
* A schedule of capital investments: if you plan to buy a rotavator or polytunnel, for example, they can go in here.

* A cash-flow forecast.
* A five-year forecast, based on all of the above.
* A one-page conclusion, drawing from all the information on the previous pages and detailing where you intend to be at the end of the five years.

This detailed planning may seem onerous if you're only planning to grow sweet peas to sell at the front gate, and you may think you have no idea where your business will be in the next year – whether, even, you'll turn out to have the intestinal fortitude to be out cutting sweet peas, whatever the weather. But, in however simple a form, a business plan will bring all your ideas into focus and give you a schedule to refer to, and is something to keep up with (or, if you're the competitive sort, to beat).

Keep accounts from day one

However small-scale you may be starting out, it's essential that you keep strict accounts. Before you buy your first packet of seed or plant your first bulb, get your accounts admin organized, otherwise there will be days when you're floundering, lost, drowning in a snowstorm of receipts and invoices. . .

If you're not confident to do it yourself, find a friendly book-keeper to help you set up a simple accounting system. Well-kept accounts will make your life infinitely easier as time goes by. As your business grows, you will have systems in place to make sure that you're up to date and tidy. The tax man is not out to get you, but if you're disorganized you may feel hounded when you have to submit a tax return.

Open a separate bank account from your personal bank account. It is easy to con yourself that you are making more than you are, or that your costs are lower than they are, if proper records aren't kept. Business is business: it involves money and a profit-and-loss account.

If you're running a business from home, you have the right to charge the percentage of your domestic overheads that is used for running the business against tax. This could include your landline, mobile phone, computer, camera, the space in your house dedicated to the business (office, storage, indoor propagating, a place to meet your customers), electricity, water rates, etc. Again, a good bookkeeper or accountant should be able to advise you on this. If you're making deliveries, you can also charge some of the running costs of your car or van – all this should be checked out.

You may not *make* a great deal of money flower farming, certainly not at the start, but you should be able to offset some of your living expenses, even if you're not yet in a position to pay yourself a salary, against the business you've begun.

Decide how much to invest

You *can* spend a fortune renting or buying land, putting in polytunnels with all-singing, all-dancing heating and watering systems; you can buy a van, a mower, a rotavator, a huge selection of perennials and shrubs for cutting – all on the back of a large loan or an investment made by yourself when you launch your business. Then you can spend on an expensive website, high-end business cards, packaging, ribbon; you may retain a public relations agency to push your business into the public eye. . .

Or you can start your business slowly – grow it, literally, one stem at a time. I believe strongly that garden farming businesses are not the sort into

If you enjoy your flower farming and find your business is growing, *then* extend your polytunnel.

We do all our floristry in what was our dining room. The office is in the sitting room, we feed people for workshops in our kitchen. . .

which massive investment should be made early on. Growing cut flowers for money is very hard work and you may not like it – so invest in what *has* worked for you, not in what you *may* not be able to make a success of. We started growing sweet peas and selling them in bunches of 20 from a barrow out front.

I remember being thrilled with £50 in our first summer (this was before I'd thought of starting a formal business). Three years after selling those first bunches of sweet peas, I began to make bouquets from my first cutting border in the vegetable garden: a 3m x 1m (10' x 3') bed stuffed to the gills with a spectacular mix of dahlias, sunflowers, cosmos, verbena bonariensis and calendula. I delivered posies to my local wine shop and began to build a small customer base.

Only then did we officially start our business and turn nearly all the vegetable garden over to cut flowers. Four years on from then and we've only just invested in our first sizeable polytunnel.

Budgeting

Keep a close eye on where you spend your money. Not only are strict accounts necessary for your business success and your sanity, but efficiently kept records are also interesting and informative, and provide a base from which to challenge yourself next year.

Your flower-farming business may not at first be the main source of income for your household – indeed, I sincerely hope that in the first year or so, while you are learning on the job, that there is another income stream on which you can depend. But unless you are keeping stringent records (of everything, not just basic accounts), you will not see where you are being successful, where you are wasting money – or whether, in fact, you're doing so well that you could drop the second income stream and trust to your farmed garden to keep you in shoe leather as well as fresh sweet peas.

Here's a simple example of costs and profit in a first season:

Costs:

❀	Seed	£200
❀	Compost, grit, seed trays, perhaps the treat of a new trowel, Vaseline to protect your seed trays from slugs. . .	£100
❀	Raffia	£25
❀	Big roll of brown kraft paper	£10
❀	Cards/advertising	£150
	Total spend:	£485

Profit:
In a five-month flowering season (April to September) you may sell 150 bouquets (that's 7.5 bouquets a week) at £15 each, making £2,250 turnover.

❀	150 x £15 bouquets in five months	£2,250
❀	Less costs listed above	-£485
	Total profit:	£1,765

This gives you £353 per month.

So, not enough to live on (nor enough to represent taxable income) – but you now have an idea of how much you need to ratchet up your planned production and sales if you do want your cut-flower patch to keep you in shoe leather. (Equally, you have an idea of what you might do over a summer in your spare time while keeping your day job. Small-scale cut-flower production is an ideal activity for somebody who can't help growing too much in the garden, and a profit of £2,250 over a five-month summer season will certainly

Hot financial tip!

From *day one* open a savings account into which you put 10 per cent of your turnover, so that you have it when you need it for your tax bill, for paying your accountant, for investment at the end of the financial year, etc. You'll then be able to reinvest the success of your business into the garden you already love. This is an efficient and sensible way to grow any business, especially one producing cut flowers.

pay for Christmas, a holiday, perhaps help with sending a kid to university. . .)

However, take off the costs of your overheads – a percentage of the mortgage/rent you spend on the land, the water you may have to pay for, etc. – not to mention the cost of your time at the minimum wage – and you may find your profits come out at zero.

But, it's your first year, and unless you have bought or rented land especially for the purpose, you would have had to pay the mortgage/rent anyway; and if you're the flower-farming type, you would have spent a great many hours working in your garden anyway, so the cost of the land and the cost to your back are relative. As I've said, this is a lifestyle choice and is unlikely to turn you into a millionaire in the foreseeable future.

If you are planning to rent or even buy land especially to start your cut-flower business, then your cost/profit balance needs to be more carefully predicted, and your business plan particularly detailed and well researched.

Late season in the cutting patch – time to take stock and spend a little profit on seed, dahlias and bulbs for flowering the next year.

Tax

In the UK you do not have to pay VAT (sales tax) until you're turning over £81,000 per year (Spring Budget 2014 level). Once you do hit the £81,000 threshold, do look at flat-rate agreements for less than 20% tax that can be made with the VAT officer for businesses which, like yours, will not be just buying in their product at the usual VAT rate and selling it on, but will be growing most of the product. Your costs will mostly be used up by water, labour (your own at first), electricity, and perhaps fuel and wear-and-tear on your car for deliveries. Your product is mostly grown from packets of seed costing £2 or so.

Having to charge 20% VAT on a product on which you've not paid an extra 20% to grow can have a horrible effect on your pricing and there-fore your relationship with your customer. If you buy in all your product and sell it straight on – for example, if you're a florist buying British-grown

flowers, arranging them in your shop, and selling them on, and if all your other bought-ledger items come in at 20% VAT – then the 20% VAT will work for you because you can claim back as much as you pay.

But if you grow most of your product, so you're only paying VAT on the price of a packet of seed, which then makes you thousands of flowers to sell on, then you might want to look at lower flat-rate VAT rates. So talk to your accountant before obediently registering for 20% VAT.

Pricing

Charge according to your experience as well as your costs. Be reasonable: don't charge so little that you can't cover your costs, but remember that you're learning on the job. In your first year you will have upsets when flowers don't condition properly, or you cut them too late and they go over, or they have a bad time in transit and aren't worth what your customer agreed they'd pay. Feel your way. When you've been going for three or four years you'll know what works for you, and you can be confident that, when you make a sale, what you supply will be better than the customer's expectations. Be honest with yourself: don't charge high-end wedding flower prices if you've never done wedding flowers before.

Speciality flowers such as peonies, roses, dahlias and even sweet peas can be sold at a premium. Look at the prices your local florists are selling at: the price you could sell *to* the florist would be something less than half that. "Less than half?!" you cry. Well, the florist will reply that they have overheads: staff, bucket-washing, their time, marketing, etc., and all that needs to be reflected in the price at which *they* sell on each flower. If your flowers are really good – high-quality, beautifully scented, fresh from the field and beautifully conditioned – don't be put off if a florist doesn't want to pay your prices. Take your product to the farm shop, the home interiors shop, or other places where they may be better received.

The price per stem

It's almost impossible to work out an exact value per stem for your flowers, especially if you're growing what you love rather than speciality flowers that can be sold at a premium. But you need to think carefully about the prices you'll sell at. If you undersell, your customers will be upset when your prices increase dramatically when you realize you're not making enough money.

If you're selling to local florists, they may want to dictate the price they pay you, but you're better off doing your maths and deciding what you will

Who will buy your stock wholesale?

Walk into a high-street florist where the staff lean boredly on a damp counter among buckets of out-of-season carnations and chrysanthemums, where you can see the swabbing bucket leaning against the back wall, and where there are bleached photographs of stiff funeral arrangements behind grizzly, week-old examples of wedding-table centres in the window. . . That florist is unlikely a) to know what to do with your flowers or b) to pay the kind of prices you want to charge.

Walk into an independent florist, where the owner has dressed the place with vintage galvanized buckets and watering cans, where the ribbon behind the counter is gingham, where there are preserving jars and interesting vases on the shelves and the owner is enthusiastically designing a client's wedding on Pinterest by photographing single stems arranged carefully on a piece of slate, and you may have found somebody who will take every flower you offer at a very reasonable price.

It's possible that your likely customer doesn't even go to your local florist, but does go to the interior design shop, the farm shop, or the café in the high street. Offer your flowers where you think they'll sell well, which is not necessarily the most obvious place.

charge. I once had a lovely chat with some Irish flower farmers who said that they were used to buyers dictating the price at which they sold their product. Don't let this happen, unless you agree that the price is fair. The price you set per stem or bunch may be the same as the florist will offer, but if you've done your calculations it means you've thought hard about your costs, your time and your overheads. Farming your garden is a lovely thing to do, but the point is to make money as well.

How to decide on a price?

Looking at a bunch of 20 sweet peas, for example, you may think, say, £3 to be a fair price. But just looking at them doesn't inform the fairness of your price. Be boring about your pricing – break down your costs:

* The price of the packet of seed – say £2.
* The time it takes to cut 200 sweet peas – say one hour at £8 (nominal value of your time).
* The time and other costs incurred in selling the flowers to your customer – telephone calls, fuel, time for pleasantries. . .
* The cost of the land rental, water, 7 months of nurturing the flowers before the first bloom in May, and so on. . .

It becomes increasingly difficult to price these sweet peas. And then there are other considerations: will your customer take a regular order of bunches of 20 sweet peas? In which case, they might get a better price. . .

In fact, at the time of writing, £3 is probably about right for *ten* stems of garden-grown sweet peas, with stems around the 20cm (8") mark. Longer stems and they'll be worth more. To get as clear an idea as possible about pricing, you need to do some research.

* Look at the sale price of your competitors' flowers. There are increasing numbers of local flower growers selling online with price lists, against which you can compare your thoughts.
* Look at the price of flowers sold in the high street. You'll probably get a third of that if your flowers are being sold on at that price – which may be fine for you, or may not.
* Then look at the amount *you* need to turn over by the end of your season in order to meet your costs and make a profit to pay yourself for your own hard work, and work backwards to come to your price. You want to sell: you don't want to price yourself so high that you end up composting your product, but equally you don't want to come out at the end of the season, having worked your socks off for demanding (perhaps very nice, but demanding nonetheless) customers, only to find you've made a loss.

Employees

Long experience has taught me that two pairs of hands do at least three people's work. So while the thought of the responsibility of employing somebody may be scary, if you want your business to grow, at some point you'll have to take the employment plunge.

There are many schemes encouraging businesspeople to take on employees, and small businesses can get tax breaks for doing so, as long as the business income is below a certain level. Look at apprentice schemes for young people, and at other organizations such as the Women's Farm and Garden Association (WFGA – which isn't just for women!), which will help you find the right people and train them. I'm not offering specific advice here, because the rules change often, but it is certainly worth investigating any help you may

Fabrizio invited an environmental health officer to visit us – who was actually very helpful, so I feel confident now that our workshop lunches meet health-and-safety criteria (as well as being delicious!).

be able to get from the government to take on some help with labour, even if it's just an apprentice.

There are, of course, certain obligations that come with taking on paid (or even voluntary) labour. If somebody works for you for more than three days a week, or for the majority of their working week, then you are obliged to offer them employment, which involves you registering with HMRC to pay their tax and National Insurance. Simple employment contracts are available on government websites.

To encourage anybody to work in horticulture is to encourage someone to work with a close understanding of the land they stand on, of the seasons as they evolve, and to learn patience and strategic planning – not to mention biology, Latin, history of art – and, in flower farming, the art of floristry. If you're like me, then you'll be proud that horticulture has made you something of an artist, biologist and botanist.

We had a 16-year-old here doing work experience for a week last year, during which she received her stellar exam results. She will probably go on to build an impressive career . . . but I know that she loved her week with us, has been teaching her mother gardening skills ever since, and has learned that you can work outside and be happy; that you don't have to sit behind a desk to earn a living. She's gladly coming back for a week's paid work during her Christmas holidays.

Insurance

If you are working from home, and you have anybody volunteering or working for you, you need insurance. It is an added cost to your overheads, but gardening can be a dangerous business, and if you're going to let anybody else free with your rotavator, or even a sharp pair of scissors, you'd be foolish not to be insured against accident or injury.

You should also have an accident book in which any accidents (small or large) are recorded, a health-and-safety document which you ask your workers and volunteers to read and sign, first-aid kits in obvious places, and somebody should be designated in charge of first aid. It's worth enquiring at your local doctor's surgery about first-aid courses you could attend.

Make sure too that your car insurance covers you for delivering your product. This ensures that you are covered in the hellish scenario of your van being stolen full of wedding flowers en route to the venue.

Environmental health

If you're feeding staff members, serving teas, holding workshops at which you offer lunch, letting people use your toilets, etc., do invite your local environmental health officer round. They will advise you on any changes you need to make to your house in order to make it 'safe'.

I nearly had heart failure when Fabrizio invited the environmental health officer in to see what we do. I immediately envisioned us having to have huge 'FIRE EXIT' signs all over the sitting room and fly-killing machines constantly *zzz-ing* in the kitchen (we live in a dairy-farming area – there are flies in the summer).

However, the officer did anything but shut us down. He was interested in what we do, and was very helpful with suggestions about ways in which we could ensure health and safety for ourselves as well as our employees. We hold lots of workshops throughout the year, and he suggested that we keep separate plates, knives and forks, etc. for workshops, that we wash potentially muddy veg in a separate sink. . . All sensible suggestions, easy to put into practice.

So it seems that local government agencies really are there to work with, not against, and while it will be against my instincts to invite government agencies into my house until the day I die, when Fabrizio does, it always turns out to be a good idea.

Planning permission

No two districts have quite the same rules regarding what structures you are or aren't allowed to put up and where. But if you're planning to put up a polytunnel of any size, do contact your local planning office and let them know what you're doing. Equally, if you're taking on a bit of land, do check whether it has an agricultural tie (i.e. can be used only as agricultural land). If you're farming flowers, you should be fine to work that agricultural land, so long as it's clear that you're not making herbaceous borders for pleasure rather than farming strips of larkspur or whatever your chosen crop for profit. Planning officers are generally helpful, and if you keep them informed, then you're less likely to find them unexpectedly standing on your doorstep questioning your plans.

Packaging and delivery

The way in which you present your product to your customer is almost as important as the product itself. It is the first sight the customer has of what they've paid for. A battered or grubby box does you no credit. You have to have a very good name to get away with using second-hand boxes.

If you're sending flowers by post, then look online for boxes. There are any number available, and I'm afraid none of them, until you're able to buy 1,000 at a time, are very cheap. But if you are building a business, part of which involves delivering your flowers far away by courier, then your initial costings are going to have to absorb those boxes. Check for solidity and for ease of opening (a lot of flowers are sent to elderly ladies).

There are off-the-peg boxes especially designed for sending flowers, with 'Fresh Flowers' written on the side. Whichever box you choose, make sure that you mark *very clearly* which is the top and which is the bottom: couriers are busy and won't stop to read small writing in thin biro. Use a thick marker – *help* the couriers deliver your parcel in the best possible condition.

Think about costing in ribbon – or raffia – or whether you'd rather just use an elastic band to separate your bunches of stems into tens or fives. You might like to have stickers made (inexpensive but very smart) to customize your boxes.

WHAT WE'VE LEARNED

We now have bespoke boxes, and I'm glad it took us a few years to drum up enough business to warrant the expense of having them designed. By the time we started thinking about what we needed in a box, we had lots of experience of good and bad boxes, having used the Post Office and then couriers to deliver. Your choices will depend on whether your flowers are being sent in or out of water, and whether they're sheaves of stems or bouquets.

Five top tips for starting a cut flower business

❀ **Make a business plan**: this will help focus your mind on what you really aim to achieve. It will help you turn a dream into reality.

❀ **Make a spreadsheet of your cut-flower year**: a patch laid out on paper will be easier to manage on the ground.

❀ **Keep accounts from day one**: systems put in place early will save a great deal of time as your business grows.

❀ **Don't over-invest at the start.** Growing cut flowers is hard work – grow your business along with your knowledge and your flowers. Whatever the size of your patch, let the business pay for your big-ticket items as a reward for sales. By then you'll know you have a reliable market for your cut flowers, and so investment will be a good idea.

❀ **Never underestimate the amount of time you should spend marketing.** If you hate the idea of having to market your product but love gardening, be a gardener.

A few words from James Cock of Flowers by Clowance

James Cock is a cut-flower grower and wholesaler from Cornwall. Flowers by Clowance is a traditional, family-run business that has survived thanks to James's sheer hard work, great charm and imaginative thinking. It is not only the high quality of his product but also his can-do attitude and ability to supply his customer, no matter what the weather, that have contributed enormously to the success of his business.

"The core of our business is growing British cut flowers on a commercial scale. Flowers by Clowance grows a wide range of cut flowers all year round under 2½ acres of modern glass, half an acre of plastic tunnels and 30 acres of arable land. We have introduced a direct-supply service via our website to meet an increasing demand for locally grown British flowers: this supplies 'home-grown / locally sourced flowers' to florists and farm shops all over the country via next-day courier. The wholesale price list goes out by email to 1,550 florists and wholesalers every week.

Our busiest time of the year has to be the Christmas period – from mid-December to the end of December. After Christmas, Mother's Day is the next most demanding for business.

Flowers by Clowance has a fleet of lorries, including two 7.5-tonne sales lorries that cover the whole of Cornwall five days a week, as well as two panel vans. The Cornwall van-sales round is a large part of the business and supplies florists, hotels and convenience stores direct with fresh wholesale cut flowers and sundries throughout the year.

My advice for sending your flowers to market? Always cut to order. Cut your flowers the day before sending, and give them a nice deep soak in fresh water, then store in a cold store to remove field heat. Always pack your flowers on the day of sending to market, to ensure they are at their best and freshest when they arrive.

Here are my five top tips for someone starting out in the cut-flower business:

* Make sure there is a market for your flowers.
* Make sure your site is level, not prone to flooding or heavy frost, and you have good windbreaks.
* Visit local florists and wholesalers. Ask what flower varieties they would like grown for them and what is not available these days. Give out free samples of your produce. Don't jump on the bandwagon and grow popular, easy-to-grow flowers – be different!
* Have a website and social media pages designed as soon as possible. These are the easiest and cheapest way of marketing your business.
* Have an understanding family: long working hours are standard."

Chapter fifteen

Where to sell

So your buckets are bursting with flowers you've cropped. Now, where are those customers? You need think about how you're going to take your crop to market, which customers you'd like to encourage, and how you're going to persuade them to buy from you. It's all very well being a good gardener, but you want to make money from your flowers too.

After all that work you've put into growing your flowers, it's important to pay as much attention to selling them. Don't just consider the obvious florist and perhaps your nearest flower wholesaler. Think also of farm shops, home interiors shops . . . perhaps the garden centre might like a cut-flower stall near their checkout? We sell a great many flowers through our local wine shop in Wincanton. Will you sell direct to customers via mail order? The list of potential outlets will be longer than the one I'm going to give you here, but, as I've said throughout this book, the idea is to get you thinking cleverly rather than tell you to do as I say.

Your local florist

Florists often complain that locally grown artisan flowers don't last, aren't well presented, and aren't reliable. It is up to you, the grower, to supply best-quality, carefully harvested and beautifully presented stock to your customer. This way, the reputation of locally grown cut flowers will improve, and all flower farmers will benefit from the efforts we make. Nobody buys from anybody just because they're local or organic: they buy from you because you're good at what you do. Being local and organic are the sugar on an already beautifully produced floral cake.

New growers can be nervous when approaching a potential customer such as a florist in their town, sometimes arriving already on the defensive, anticipating a brush-off. So, take some time to ensure that you impress your local florist: be professional.

* Arrive with bunches all cut to the same length, keeping the stems as long as possible, neatly tied at two places on the stem with raffia (so that the stems don't separate and get caught up with each other in the bucket).

* Keep your flowers in water while you deliver so that they don't have time to faint and need rehydrating after the journey in a hot car.
* Deliver your flowers in a clean tin bucket: one the florist can use in the shop.
* Tell the florist the (reasonable) price you want per stem – don't wait to be told what they will pay. Farmers are used to being told the value of what they grow, but you should work out what to charge and decide what your crop is worth. If you sell your flowers at much less than you think they are worth, you'll make a loss, hate what you do, and give up.
* Tell them how long you expect your flowers to last. Offer a free sample so that they can see that you're telling the truth.
* Tell them also whether you've treated your flowers with flower-preserving liquid, or with any chemicals which they might otherwise duplicate by treating the flowers again. If your flowers are untreated by chemicals, explain that they will, therefore, grow and develop in the vase.

You will have to *sell* your product to local florists – you're unlikely to find them falling over themselves to buy your flowers. So, tell them about your patch, what else you grow. Give them a story to sell to their customer – the story of artisan-grown flowers: feeding the bees, costing very little in flower miles, free of chemicals, eco-friendly packaging (a reusable tin bucket and a couple of strips of raffia has to count as minimal packaging!) Make it easy for the florist to sell your flowers on. If you don't value your product, the person you're selling to won't value it either.

In the UK, high-street florists have a very strong relationship with their (often Dutch) suppliers. The florist can order online at 10 p.m. for delivery by 8 p.m. The stem prices are affordable, and while the florist's current wholesaler may not be able to

Nobody buys from anybody just because they're local or organic: they buy from you because you're good at what you do. Being local and organic are the sugar on an already beautifully produced floral cake.

offer the variety or freshness of flowers that you can offer, the wholesaler will supply the florist with most of what they need year-round.

You will have to persuade your local florist that you can compete on price and quality, and certainly on variety and seasonality, with the wholesaler with whom they already have an established relationship. Florists don't make a huge amount of money. The fact that they may be able to pay

their wholesaler on a 60-day basis may mean that they simply can't afford to buy from you.

So, this will not be an easy relationship to build, and you may find that for retail outlets it's easier for you to sell to or through farm shops or gardenalia and kitchenalia shops – perhaps even through cafés, wine shops, home interior shops or even bookshops – than try to persuade your local florist to sell your wares. Remember too that any retail

Posies boxed up for sale at our local wine shop, which does a roaring trade in our home-grown flowers.

outlet will need to add at least a third to the price you're charging (though they'd probably rather double your stem price) to cover their overheads.

Farmers' markets

Farmers' markets are a fairly obvious place to start selling for any flower grower. They are usually welcoming to people supplying locally grown produce, and while there *is* a growing network of small-scale cut-flower producers, you will probably not be treading on any toes if you call your local farmers' market and ask if you can have a stall.

Do think a bit about the people who will buy at *your* farmers' market. If you're selling in a tiny town where people go for chore shopping rather than luxuries, then you won't do well if you try to sell bouquets at a premium. At that kind of farmers' market you'll do better selling flowers by five, ten or fifteen stems at a time, at a reasonable price.

Offer your customers appealing ideas for how they might arrange your flowers at home.

If, however, you can get a stall at a market in a bigger town, where people go for time-off shopping, with money in their pockets to spend on luxuries, then you could think about selling bouquets and posies at more of a premium. Perhaps you could experiment at two different farmers' markets and see which suits you best.

Remember that while farmers' markets feel like a great place on principle, there are costs involved: your stall, transporting the flowers, perhaps a banner telling your customers who you are and where they can find you on non-market days, and business cards or leaflets to give your customers.

Farmers' markets are also a great way to tell potential customers in your locality that you're available for more than that Thursday-morning slot. Have leaflets to hand about wedding flowers and pick-your-own days, and information about supplying local shops and cafés – be prepared to use the venue to sell more than just flowers by the stem.

Gate sales

This is a great way to sell flowers for the beginner grower. You can make a stall or a stand at your front gate. You can choose what price you sell your flowers for. There are no costs involved. You don't even have to leave the house. So long as there's somewhere safe for cars to stop and buy, then you can test your wares, test your prices, and test your floristry skills on your passing trade.

If, however, your gate is very out of the way, and there isn't much in the way of passing trade, you may want to think again. Equally, if your gate is exposed so that the flowers are in glaring sun all day, or thrashed by the wind and the rain, then you might want to think about giving them some protection. Needless to say, make sure you have somewhere safe for your passers-by to stop. If you live on a fast, dangerous road you might have to reconsider gate sales.

Pick-your-own

Many small flower farms have pick-your-own sections, or pick-your-own days scheduled throughout their season. If you're likely to have a glut of something – sweet peas or dahlias, for example – then a pick-your-own day is a great way to get a lot of stock cleared from your patch.

Equally, if you're growing in a really small way but like the idea of raising money for charity or a local cause, then a pick-your-own day with teas and cakes sold on the side is a great way to raise money. Perhaps you already open your garden for local charities or the National Gardens Scheme? Incorporating a pick-your-own cut flowers option in your open days will increase footfall and widen your potential customer base.

Make sure you've thought about all the implications if you're going to allow people on to your land to pick. It's worth asking someone in to make

a risk assessment (we live in a litigious age!) and checking that your insurance covers visitors walking on uneven ground carrying scissors – whether your scissors or theirs!

So long as your domestic kitchen is clean, you won't need special health and safety arrangements for serving teas and cakes, but make sure you do your research about legal responsibilities before selling anything edible from your house.

While a pick-your-own day is a great way to use up stock, and even a way to run your business – you could have pick-your-own day every Thursday and Friday throughout the season, for example – think hard about how you're going to charge people who come through the gate.

* You could have a flat fee: £10 per bucket, which people are given as they arrive and they can fill with as much as they can fit into it. You supply the buckets, so control the size and potential number of stems.

WHAT WE'VE LEARNED

Make sure your honesty box, if you have one, is secured. The sort of people who stop and buy flowers from roadside stalls aren't generally the thieving kind, but there will be other people passing who can't resist an unlocked honesty box, or one they can just pick up and carry away.

Local success

A farm near here has given over a single field to growing flowers, and they open from May to September with scissors and string available in the shed and an honesty box with a sign reading "Pay what you think your flowers are worth". The gardeners are sometimes there; sometimes not. They know that the honesty box will be filled by the kind of customers they get. They wouldn't still be using this arrangment, ten years on, if they weren't making enough from their pick-your-own system for it to be worth their efforts.

You can offer mixed buckets of flowers for people wanting to do their own wedding flowers.

- Or you can ask that people pay a 'contribution' according to what they think their flowers are worth – this is often a way to make more money than you would if you charged a flat fee. People who have a lovely time at your plot, enjoying the sunshine, pottering about, living your dream for an hour or so, may pay you a great deal more for their bucket of flowers if you let them make a contribution than if you charge a specific amount. They'll pay you for the sunshine, the view and the lovely after-noon outdoors, as well as for the flowers they take home.
- If you charge per stem, or per ten stems, for pick-your-own, then you increase the admin involved and need to have somebody prepared to count up and charge at the exit. It just depends on how complicated you want your system to be.

Wedding fairs

If you'd like to sell the flowers you grow to brides – whether you're going to do the floristry yourself or would like to encourage do-it-yourself brides to buy their flowers from you – wedding fairs can be a good way to get your name out there.

Choose your fair carefully: some wedding fairs are large, corporate events at which suppliers elbow their way to the front of the crowd to shove their leaflets in the hands of rather stunned-looking brides-to-be. You don't want to spend the day trying to sell home-grown flowers to brides who want blinding pink gerberas in goldfish bowls – you're wasting your time and their time that way.

Home-grown flowers work well at vintage-style wedding fairs: fairs where you might be able to join forces with someone hiring out vintage china, for example. This arrangement will reduce the cost of your day and give you someone enjoyable to spend the day with, and customers already sympathetic to your product are more likely to be drawn by the extra dimension to your stall.

Before you book your place, ask the wedding fair organizer a few questions:

* How many visitors do they expect on the day?
* If they've held the fair before, what was the footfall that day?
* How many other flower growers or florists will there be there? Too many other florists and you'll struggle to be seen unless you can make a real splash – which costs money!
* Where will your stand be placed? Make sure that you won't be tucked away at the back of the room.

Do have on your stand:

* A great many leaflets. (We also tend to let other people use our flowers to dress their stand in return for having our leaflets on display.)
* A rough guide to pricing.
* Lots and lots of photographs. Wedding fairs often happen out of season, when you may struggle to have stock with which to demonstrate your skills – but a well-set-up stand with photographs, a big banner, a lot of information about your plot and what you do, and perhaps even a screen with rolling images on it, will draw the brides in.
* Be prepared to get out there and talk to the brides, rather than waiting for them to come to you. Wedding fairs are extremely competitive arenas, and if you sit demurely behind your stand waiting for people to talk to you, you risk being ignored altogether.

Five top tips for selling flowers

* Remember that nobody will buy from you just because you're eco or local: they'll buy because your product is high quality and something they can't get elsewhere.

* Deliver on your promise. Sometimes it's easier to make the sale than to follow it through. Flowers are as temperamental as the weather. Hedge your bets with planting, set up emergency resources (develop a good enough relationship with your nearest competitor that they might supply you if you need to supplement your stock), and you will be able to deliver what you've promised.

* Photograph everything. If you're doing wedding fairs in winter when you have no stock with which to show off your skills, make sure you've got some wonderful photographs with which to inspire potential brides. Have photographs to blog with, to keep your website fresh, and to be able to send as stock pictures to your local press if they want to do a story on you.

* Give good advice to your customer on how to care for your flowers. If you're not treating the flowers with chemical flower food, say so. Tell your customer what to expect from your flowers and how to get the best out of them.

* Remember that part of your story is that you're local; that the flowers you grow help support the wildlife in your locality – that they feed the bees. People are buying a little rural dream, and while your reality may be mud and manure and blisteringly early starts in the morning, you should sell them the dream – an orchard, apple blossom, tulips. . .

Chapter sixteen

Marketing & social media

Build a social media platform – or opt for traditional leaflet drops and advertising. Give talks to local horticultural societies. Open your garden for people to come and see what you do. There are so many different ways to tell potential customers about your flowers and to encourage them to buy – but tell them you must.

Remember at all times that tea for the workers makes things run smoothly. In fact, this would make a great picture on Instagram or Twitter – engaging with people about what a day on a flower farm is like.

Running any business is 20 per cent about producing the product you want to sell, and 80 per cent about marketing. It takes a great deal more time to sell cut flowers than it does to grow them.

There are, of course, all the traditional avenues: local press, radio, advertising and so on. But it seems to me that calling local journalists, having a chat, impressing upon them the value of their coming to look at your plot, etc. is time-consuming and potentially expensive. For the price of a good mobile phone you can set up Facebook, Twitter, Pinterest and Instagram accounts and start telling the story of your growing patch from day one: all from your phone, all on the hoof, taking no more time than needed for the click of a photograph and the whoosh of a send.

I will admit to being a bit of a social media bore, but we wouldn't have a business at all without Twitter. Where we live is extremely rural, and in our local town we just wouldn't get the footfall to keep a family of four on bunches of flowers. Social media has given us a worldwide platform, and thanks to that technology we have customers in Perth (that's Perth, Australia), San Francisco and LA, as well as all over the UK and Europe.

So from day one you cannot do better than get your business profile online, and start using it to tell your friends, your family, and all the other people who will join in if you're engaging enough, about what you're doing.

Social media

The Internet has made it incredibly easy for small businesses to connect directly with their potential market without advertising or costly PR.

The Common Farm Flowers social media story

We have built our business entirely on the back of social media interaction. Here's a précis of how and why we did it.

When we had been in business for about ten months, we were quoted a retainer fee of £10,000 a year for the services of a very good, and well thought of, local PR agency. This quote came after their representative had driven over to us from their office half an hour away, spent a couple of hours discussing our business, driven back to their office, done quite a lot of other work, and a week later emailed us a proposal.

During that week I'd spent half an hour a day on Twitter and Facebook and had increased our following on Twitter by 100. It occurred to me that if we spent £10,000 retaining this very well-respected agency, there would be quite a lot of driving to and from offices, long hours spent in meetings briefing ideas, waiting for emails and going over results. They said they'd certainly be able to get us into our local magazines and that there was a possibility they might be able to get us some national press.

For the possibility of some national press we were being invited to spend not only £10,000 a year but also, I calculated, about five whole working days in that year to-ing and fro-ing – both physically and through emails, signing off specs, etc.

Time is valuable. I decided I'd commit to continue spending half an hour a day on social media, save that £10,000 (which the business didn't have to spend at that stage anyway) and see if we could get some press that way. Since then we've been featured in *Country Living*, *Good Housekeeping*, the *Telegraph*, *LandLove* magazine and *The English Garden* magazine, and are often mentioned in the gardening or home sections of the weekend broadsheets – all because of Twitter. Please don't think I'm just showing off: I want you to see what you can do for free – with a little tenacity – to publicize your work yourself.

Twitter especially has been our marketplace, our press contact book, and our support system. . . I cannot recommend more highly using social media to raise the profile of your business and what you do. It is free. You control what you say or do on it. You will, if you use it wisely, make direct contact with your potential customer base, and all the press you need.

Pitfalls to watch out for

Remember, even if you're chatting with friends, that Twitter, Facebook or whichever other platform you're posting on reflects your brand. However small your cut-flower patch, it is your business, so remember to tweet or post about what your product is, what your garden looks like, how your customers can find you, and so on.

As with dinner parties, avoid discussing religion or politics – whichever way you lean, you'll offend one customer or another. Similarly, don't swear on social media. Would you swear in conversation with a customer in your shop? Then think of social media as your shop and avoid explosive language. And, assuming that you are not a breathy teenager, avoid OMG-style comments.

Never post on social media sites after a glass or two of wine (gin, absinthe. . .) You'll only find yourself breaking all the above rules and regretting it all in the morning.

Top social media platforms to choose from

There are so many different platforms for using social media. Choose a couple to focus on.

Pinterest is great if you aren't wordy but like taking pictures: make sure the pictures you pin of your own work come from your website or blog (see overleaf), or the links back from Pinterest will just say 'pinned by pinner' and you won't get the potential search engine optimization you could have had for your website.

Instagram is incredibly popular for people who take pictures on their phone.

Twitter is brilliant for those who can turn a sharp phrase – plus, *all* the press you need is there for you to find.

Facebook is boring, slow and changes the parameters all the time, *but* more people use Facebook than any other social media platform, so it's worth posting once a day.

You should also keep a blog, which people can follow to see the week-to-week development of your patch. Three paragraphs and two photographs make a nice little blog post. This will not only tell your potential customer base what you're up to, where they can find you, etc., but will also give you a wonderful diary history of your flower-growing business. Blogs are free to set up from Blogger or Wordpress.

Website

All this social media interaction helps drive your search engine optimization, to keep your website popping up on page one of search engines. So make sure you have a website! A website reassures people that you are a bona fide business. It is like a freely available online business card. It needn't be complicated, difficult to manage, nor cost a fortune. You can design a good, simple site yourself, using a blog such as Blogger or Wordpress, or have one designed for you. If you decide to pay for a website, make sure that you will be able to update it yourself easily and, if it is built separately from your blog, that your blog is embedded *in* your website. Ask your designer to make sure that your web stats are easily available to you, so that you can see how your website is doing.

At first I would keep it very simple: a home page with a picture of you and what you do, and contact details with a weekly blog post. Even the simplest sites can be given add-ons as time goes by; you can introduce new pages for new products as and when you expand. You might list the dates you'll be at which farmers' market, for example, and which wedding fairs you plan to attend.

I can feel some of you looking at me with despair, saying, "I just want to grow cut flowers!" Yes, but you also want to sell them, and how else are you going to find your market?

Local press

If you find the idea of spending a little time every day posting pictures of what you're up to on your Twitter and Facebook accounts too daunting, then you could give yourself confidence by being old-fashioned in your approach and simply picking up the phone.

Your local papers, local radio and local advertisers are always looking for content, and a new business doing something interesting and charming like yours makes great copy. Give them the story: why are you starting your business? Are you changing career? Do you want to save the bees? Have you thrown in the towel after 20 years as a lawyer to devote your life to your garden? Were you made

A cut-flower plot is a photogenic story for your local press.

You and your helpers should wear your brand name all the time while working, so people remember who you are.

Leaflets, as well as samples of your flowers, left in shops and cafés will help, but beware leaving them too long (especially in the case of flowers), so that they don't wilt, or fade, or are full of last year's news. We still, years since we started, keep

redundant and are you spending your redundancy package on a polytunnel and a shrub collection? Are you taking up flower farming having retired? Ring the news desk, tell them what you're doing, and invite them along to have a look.

Host a press event when your first-season flowers are just beginning to look fabulous. It won't cost you very much to lay on cups of tea or glasses of wine for your local press one evening, while the sweet peas dance in the wind and the calendula smile up at them. Flower farms look especially lovely in the summer, and yours is a very pictorial story for your local press to tell.

Where else to tell your story

You also need to tell your potential local customer base that you're growing flowers, remembering that not *everyone* is plugged in to the Internet.

Copyright

If your cut-flower patch is beginning to make you some money and you are perhaps thinking of expanding and investing in some stylish branding, then do look at intellectual copyright. It would be a shame if, once you are really making a name for your business and have lovely branding designed, branded boxes made and ribbon printed, you found that somebody else had copyrighted the name and were trading under it elsewhere – or, worse, had copyrighted it and wanted you to buy it from them for a fat fee.

In the same vein, if you post photographs to demonstrate what you do on Twitter, Facebook, your blog and elsewhere, then do add a watermark or a © to them before uploading them. We resize our photos to a size which still looks good on our Twitter feed, but which makes them not worth copying by anyone else. Mention at the head of your blog, Pinterest and Facebook pages that you claim copyright of all pictures you post on there. Similarly, if you post other people's pictures, then acknowledge their copyright or at least say whose pictures they are.

There are millions of pictures posted on all kinds of social media platforms every day, and it's unlikely that anybody will be interested in purloining yours and claiming them as their own – but it's worth making sure they don't.

our local wine shop filled with fresh flowers and our business cards, and over the years we've had weddings, bouquet commissions and funeral orders from regulars who like our flowers and see them change, week after week.

Call your local horticultural societies and gardening clubs and offer to give a talk. Take flowers along and demonstrate what you do. Take your business cards, and hand them out at the end.

Join a local networking group. Networking groups are to local-business-building what social media is to a wider audience. They are a great way to find out what other businesses are up to and what they might like you to do for them. Over breakfast once a month (or elevenses, or lunch, or tea, or supper – depends on your group), a selection of people get together and talk about what they do. You might find yourself supplying weekly flowers to an accountant's office, or designing a hair-and-flowers package with a local hairdresser for weddings and events.

Like social media communities, networking groups may at first seem made up of too random a group of individuals with little in common – but people are more likely to do business with people they know, so be the person your business colleagues in your local town know and they'll come to you, and will pass on your name to their own contacts.

Five top tips for marketing your product

❁   Remember: running a business is 80 per cent marketing and 20 per cent producing the product you want to sell. You need to *sell* your flowers as well as grow them.

Call the local press and radio stations and tell them what you're up to: flower farms make very attractive features when they're in full flower.

❁   Only engage with social media if you intend to keep it up: one tweet every six months will not reflect well on your business, and people might think you've given up if you only tweet once then stop. One tweet a day shows that you're engaging with your customers.

❁   Follow other people in your industry to see how they tweet (I'm @TheFlowerFarmer), how often, and what they tweet about.

❁   Follow your industry press: gardening, journalists with an eco bent, 'Best of British'-type press, home interior and lifestyle magazines. Tell them what you're up to. Follow also your local farmers' market and any feeds promoting businesses working in your county or town. Tell them what you're up to, engage with them, and they'll tell their followers what you're up to too.

❁   Join a local networking group to pick up business leads you might never have come across otherwise.

Afterword

Some people have said I'm mad to have written this book. "Why encourage others to try and share your market?" they ask. "Surely you could be *the* flower farmer and have the British flower market to yourself?" But no matter how lofty my ambitions, I'm not sure I can service a £2.2-billion industry single-handed. Besides, I hope this book will reach far beyond the UK, to anyone who might be thinking that their cut-flower patch could be made profitable.

And in the same way that if there's one antiques shop in a high street, you may drive on through, but if there are ten you may well stop to have a wander, I'm convinced that the greater the number of people growing cut flowers in a local, sustainable, artisan way, the greater the market for those flowers will become.

Let us give our customers the opportunity to choose high-quality, seasonal flowers, grown by people who by law are paid a living wage and prevented from wasting or poisoning their land's resources: flowers that are fresher by days than those flown in from abroad.

Cut flowers in any house are a gorgeous luxury: a rose, a sweet pea, a dahlia – none of these are going to feed a family, but everyone likes a little treat now and again. Let us supply our neighbours, friends and new-found band of customers with a luxury they can be proud of. Let us create a product of such high quality that everybody who buys cut flowers for a present, a thank you, for congratulations, for commiserations, in remembrance . . . assumes that the best flowers they can buy will be grown relatively locally. Let us show our customers day by day that seasonal, locally grown flowers are beautiful, scented, and totally different from those offered by far-off competition.

So this book is, in a way, not so much a gauntlet laid down but a load of gardening gloves scattered for those who'd like to pick them up and dig for this challenge. Let us rebuild the locally grown cut-flower market around the world – from the earth up. Let us remind people that we can grow flowers year-round to compete successfully with any product from afar. Let's see a bunch of locally grown cut flowers on every kitchen table, from Arkansas to Adelaide to Adlestrop.

Appendix 1:
The flower farmer's year planner

The flower farmer's year is always busy, so it's worth planning it out, best on a spreadsheet, to make sure you don't miss out any important jobs. The most important thing to remember of all, however, is why you want to grow flowers – and, whether it's for pleasure, profit or both, to make time to stop and admire your handiwork, breathe in the blissful scent of that first sweet pea when it flowers, take photographs of your massed ammi plants dancing in the breeze to cheer yourself up in the black-dog days of winter, and enjoy the fact that a productive cut-flower patch will give pleasure to so many more people than just yourself.

The following table shows the jobs to be done at each time of year here at Common Farm. For readers in other parts of the UK, or other parts of the world, I have included 'translations' of each month, so that you can adjust your planning to suit the climate and seasons where you live.

THE FLOWER FARMER'S YEAR				
	Sow/plant	**Harvest**	**Propagate**	**Other**
September (Early autumn)	Plant narcissi bulbs. Sow first crop of hardy annuals. Plant biennial seedlings in flowering positions. Plant out perennial seedlings.	Dahlias End-of-season annuals Acidanthera Hydrangeas	Take cuttings of lavender and tender herbs. Take soft and semi-hardwood cuttings of shrubs – these can root over winter for potting on in the spring. Split spring-flowering perennials to increase stock. Chuck any spring-flowering stock which didn't perform well enough for you, or which you didn't like. Take hardwood and semi-hardwood cuttings.	Clear hardy annual beds, and mulch and plant up where your plan dictates. If leaving beds fallow over winter, you might direct-sow a green manure to suppress weeds. Order bare-root shrubs and roses for planting in winter when dormant
October (Mid autumn)	Plant alliums.	Chrysanthemums Dahlias Nerine lilies Schizostylis		Move tender perennials to overwinter under cover. Mulch empty beds. A good layer of well-rotted manure on empty beds now will hopefully be broken up by frost for raking in next spring.

THE FLOWER FARMER'S YEAR				
	Sow/plant	**Harvest**	**Propagate**	**Other**
November (Late autumn)	Plant tulips. Plant bare-root shrubs and roses if dormancy has begun.	Chrysanthemums Nerine lilies Schizostylis		First prune of roses when dormancy begins. If your polytunnel is clear, plant out hardy annual seedlings to overwinter there and give you an early spring crop. Lift dahlias and store after first frost.
December (Early winter)	Plant bare-root shrubs and roses.	Dogwood stems for colour First narcissi Forced bulbs for the Christmas market Willow stems for colour Winter foliage		
January (Mid winter)	Sow second crop of hardy annuals, under cover.	Dogwood stems for colour First hellebores Forced winter bulbs Hazel catkins Narcissi Willow stems for colour and pussy willow Winter foliage		
February (Late winter)	Sow third crop of hardy annuals, under cover.	Forced winter bulbs Hellebores Narcissi Snowdrops Winter-flowering euphorbia Winter foliage	As you start digging over borders, split late-summer-flowering and autumn-flowering perennials to increase your plant stock.	Get rid of any perennials which under-performed or weren't profitable the previous season. Prepare beds for direct-sowing. Top up empty-looking raised beds with new compost. Prune roses again, and buddlejas. Prune fruit trees and bring in prunings to force.

THE FLOWER FARMER'S YEAR			
	Sow/plant	Harvest	Propagate	Other
March (Early spring)	Sow half-hardy annuals under cover. Direct-sow hardy annuals once the ground warms up. Plant summer-flowering bulbs. Prick out and pot on seedlings as necessary – you can tell it's time to do this if you can see a root curling out of the bottom of a pot or a tray.	Anemones grown under cover Ranunculus grown under cover Early-flowering narcissi and daffodils Forced blossom from prunings		Order biennial and perennial seed to sow in June. Order late-summer-flowering and autumn-flowering bulbs. Make a batch of compost tea and feed soil weekly throughout the summer.
April (Mid spring)	Pot up dahlias and put under cover to sprout. Direct-sow last crop of hardy annuals. Sow tender annuals under cover. Plant autumn-flowering bulbs. Keep pricking out and potting on seedlings as the season progresses.	Anemones Cowslips First cow parsley First tulips Forget-me-nots Iris Narcissi and daffodils Ranunculus Spiraea – blossom and new leaf, various varieties	Check overwintered cuttings for roots and pot on if necessary.	Make nettle tea.
May (Late spring)	Pot on dahlias if necessary – hold them back before planting out if there is the slightest risk of frost. Plant out all remaining seedlings, remembering still to harden off before planting out – a late frost will kill tender seedlings.	Lilac Alliums Cow parsley Cowslips Forget-me-nots Peonies Red campion Sweet peas Sweet rocket Sweet Williams Tulips		Feed plants with nettle tea.

	Sow/plant	Harvest	Propagate	Other
June (Early summer)	Sow biennial and perennial seed. Don't forget to water seedlings and pot them on as they grow, to make good young plants for planting out in September. Plant out dahlias (1 June).	Cornucopia of flowering annuals Alliums Delphiniums Roses Summer-leafing foliage Sweet peas		Clear and lift spring bulbs if you're treating them as annuals. Feed beds where the bulbs were. Use the space to plant dahlias, perhaps, or tender annuals. Feed plants with nettle tea. Make comfrey tea.
July (Mid summer)		Second crop of hardy annuals First half-hardy annuals Gladioli Larkspur Roses Summer-leafing foliage		Feed plants with comfrey tea to encourage flowering. Remember your biennial seedlings – water, prick out, and pot on if necessary.
August (Late summer)	Order spring-flowering bulbs: remember to stipulate when you would like your order to be delivered. Order fresh hardy annuals seed for September planting.	Tender annuals – amaranthus, sunflowers, zinnias Dahlias Hydrangeas Sunflowers		Clear spent crops, mulch beds and prepare for planting with biennial seedlings.

Table title: THE FLOWER FARMER'S YEAR

What we grow at Common Farm

We grow a great many varieties of flowers and shrubs for cutting here at Common Farm. We are ecologically minded; our flowers are grown in small chunks, sown successively: there is never *masses* of anything, but a constantly changing palette of colour and texture, which I hope makes each one of our bouquets unique, as well as completely different from anything you could find anywhere else. Even if you were growing exactly the same list of plants and flowers as us, I'm sure they wouldn't be put together in the same way – you'll be driven by *your* taste, *your* experience, the colour of the sky at *your* studio that day. . .

This is last year's list. No doubt next year's will be different. The wonderful thing about growing as many annuals and biennials as you do when growing cut flowers is that you're not stuck with any particular variety, and can be as unfaithful as you like – diving into a love of a newly discovered victim of your itchy-scissor syndrome just because the idea tickles you. For the price of a packet of seed, how can you resist an experiment? Although, of course, space is always an issue, and you may have no extra room in your plan. . . Growing cut flowers for money is a job for the mildly compulsive character. Don't say I didn't warn you!

I'm not giving varieties here, or the list would be too vast. This is just a snapshot, to give you an idea of what you might get in a bouquet grown, cut and tied here at Common Farm in Somerset at any given time of the year – and perhaps it might give you some ideas of what *you* might grow. Please remember that our soil is thick clay, our conditions are generally wet (West Country wet), and we garden in raised beds. In the winter we do supplement our stock with flowers from Cornwall and Lincolnshire, but we only ever use British-grown flowers in our bouquets.

There is almost nothing at Common Farm that is safe from my cutting scissors. I will experiment with everything from Scots pine in winter to hawthorn blossom in summer. So far I have found only acanthus leaves to refuse resolutely to condition, and so I have gleaming cushions of them in a garden that otherwise looks as though it suffers from a very bad hairdresser.

But – we're growing flowers for money – and we're beginning to make a relatively good living from it.

Spring harvest

Annuals
Calendula
Cerinthe
Cornflowers
Orlaya
Sweet peas

Biennials
Foxgloves (maybe four or five kinds)
Honesty (purple and white)
Sweet rocket (purple and white)
Sweet William (lots of different varieties)
California poppies (several varieties)
Big splashy poppies (a few – they're difficult to cut and condition)

Bulbs/corms
Anemones (in the polytunnel)
Alliums (four kinds)
Camassia
Narcissi (five or six varieties)
Ranunculus (in the polytunnel)
Tulips (ten varieties, or more – depends on mood and cash flow when ordering)

Perennials
Aquilegia (blues, pinks, purples, whites – they interbreed: I can't control them, let alone name them)
Bistort
Euphorbias (especially sun spurge)
Peonies (just a few, all different, very lovely)
Tellima

Herbs
Parsley (cuts well when it flowers)

Wildflowers
Buttercups
Cowslips
Dogwood
Guelder rose
Forget-me-nots
Red campion

Shrubs and trees
Elder
Hornbeam
Lilac (two varieties)
Oak
Snowball bush
Spiraea (three varieties)

Fruit blossom
Apple, pear, plum, cherry (several different varieties of each)

Summer harvest

Annuals

Calendula (two or three different kinds, plus the rogues that spring up all over the garden)

Annual chrysanthemums (four or five varieties – a big favourite of mine)

Cornflowers (four or five different colours)

Cosmos (two or three kinds)

Larkspur (five or six colours)

Nigella (faithful old 'Miss Jekyll')

Phacelia

Rudbeckia

Snapdragons (three kinds this year)

Sweet peas (maybe 20 varieties – again, depends on mood, and on colour trends, wedding bookings . . .)

Perennials

Agapanthus (two varieties)

Alchemilla

Astrantia

Clematis (I can see we will make a good collection of clematis – it's wonderful cut year-round)

Dahlias (30 or 40 varieties, depending on how well they over-wintered and how much fresh stock I order in)

Delphiniums (two sorts – so far)

Honeysuckle (I have a growing collection)

Loosestrife 'Firecracker'

Pinks (about five different ones – I love the scent)

Roses (maybe twelve kinds)

Scabious (several – maybe four varieties)

Veronica (so far, one – so far. . .)

Herbs

Catmint

Feverfew

Lavenders

Lemon verbena

Marjoram

Melissa

Mints

Sages

Thymes

Wildflowers

Devil's bit scabious

Knapweed

Meadowsweet

Ox-eye daisies

Shrubs and trees

Philadelphus

Physocarpus (two kinds)

Smoke tree

Autumn harvest

Annuals

Bells of Ireland

Sunflowers (two or three varieties – I grow these mostly for fun)

Wild carrot

Zinnias

Perennials

Asters (the bullying blue perennial one, which is very useful)

Bronze fennel

Chrysanthemums (a few, but I feel I may become more enthusiastic about these)

Crocosmia (two kinds)

Persicaria milletii

Sedums (three or four different colours)

Schizostylis

Verbena bonariensis

Herbs

Garlic chives

Shrubs and trees

Buddleja (probably five different kinds)

Winter harvest

Blackthorn (sloes)

Brachyglottis

Euonymus (two kinds)

Euphorbia

Ivy

Old man's beard

Rosemary

Willow

Appendix 2: Plant names

The following table gives the common and Latin names for the plants mentioned in this book. The first column gives the name by which a plant is commonly known to gardeners: this is often an 'anglicized' or abbreviated version of the Latin name, rather than the plant's traditional common name.

Commonly known as	Latin name	Also known as	Notes
Acidanthera	*Gladiolus murielae*	Abyssinian gladiolus	
Achillea	*Achillea* spp.	Yarrow	In the wild this is known as yarrow.
Agapanthus	*Agapanthus* spp.	African lily	
Alchemilla	*Alchemilla mollis*	Lady's mantle	
Allium	*Allium* spp.	Ornamental onion	Lots of choices for great cut flowers.
Amaranthus	*Amaranthus caudatus*	Love-lies-bleeding; careless	
Amaryllis	*Hippeastrum* spp.		
Ammi	*Ammi majus*	Bullwort	Many varieties to choose from.
Ammi	*Ammi visnaga*	Toothpick bishop's weed	
Anemone	*Anemone cornaria*		We grow the De Caen Group.
Angelica	*Angelica archangelica*		The green angelica, which we grow.
Apple	*Malus* spp.		
Aquilegia	*Aquilegia* spp.	Columbine; granny's bonnet	
Artemisia	*Artemisia ludoviciana*	Western mugwort	
Aster (annual)	*Callistephus chinensis*	China aster	
Aster (perennial)	*Aster* spp.	Michaelmas daisy	
Astrantia	*Astrantia major*	Masterwort	
Bay	*Laurus nobilis*		
Beech	*Fagus sylvatica*	Common beech	
Bells of Ireland	*Moluccella laevis*	Irish bellflower	
Bistort	*Persicaria bistorta*	Common bistort	
Blackberry	*Rubus fruticosus*		Keep gauntlets for cutting this!
Blackthorn	*Prunus spinosa*	Sloe	The sloes are an added extra!
Bladder campion	*Silene vulgaris*	Common bladder catchfly	
Bleeding heart	*Dicentra spectabilis*		
Bluebell	*Hyacinthoides non-scripta*		This is the bluebell native to the UK.
Box	*Buxus sempervirens*		
Brachyglottis	*Braychyglottis* (Dunedin Group)		
Brodiaea	*Brodiaea elegans*		
Bronze fennel	*Foeniculum vulgare* 'Purpureum'		

Commonly known as	Latin name	Also known as	Notes
Buddleja	*Buddleja davidii*	Butterfly bush	
Bupleurum	*Bupleurum rotundifolium*	Hare's ear; thorow-wax	
Buttercup (meadow)	*Ranunculus acris*		
California poppy	*Eschscholzia californica*		
Camassia	*Camassia leichtlinii*		
Campanula	*Campanula medium*	Canterbury bells	
Catmint	*Nepeta grandiflora*		
Cerinthe	*Cerinthe major* 'Purpurascens'	Honeywort	
Cherry	*Prunus* spp.		Cut after leaves out, to avoid leaf curl.
Chives	*Allium schoenoprasum*		
Choisya	*Choisya ternata*	Mexican orange blossom	Some people hate the smell. . .
Christmas box	*Sarcococca* spp.	Sweet box	
Chrysanthemum	*Chrysanthemum* spp.		
Chrysanthemum (annual)	*Chrysanthemum carinatum*		
Clary sage	*Salvia viridis*	Annual clary	
Clematis	*Clematis* spp.		
Cleome	*Cleome hassleriana*	Spider flower	Some cut-flower growers swear by it!
Comfrey	*Symphytum officianale*	Common comfrey	
Coriander	*Coriandrum sativum*	Cilantro	
Corkscrew hazel	*Corylus avellana* 'Contorta'	Harry Lauder's walking stick	
Corkscrew willow	*Salix babylonica var. pekinensis* 'Tortuosa'	Dragon's claw willow	
Cornflower	*Centaurea cyanus*	Bachelor's buttons	Good also for confetti.
Cosmos	*Cosmos bipinnatus*	Mexican aster	
Couch grass	*Elymus repens*		Bane of my life. . .
Cow parsley	*Anthriscus sylvestris*		
Cowslip	*Primula veris*	Bedlam cowslip	
Crab apple	*Malus sylvestris*		My favourite fruit to grow and cut.
Crack willow	*Salix fragilis*	Brittle willow	
Craspedia	*Craspedia* spp.	Billy buttons	
Creeping buttercup	*Ranunculus repens*		
Crocosmia	*Crocosmia* spp.	Montbretia	
Crocus	*Crocus* spp.		
Daffodil	*Narcissus* spp.		
Dahlia	*Dahlia* spp.		
Delphinium (perennial)	*Delphinium* spp.		We grow the Pacific Giants.

Commonly known as	Latin name	Also known as	Notes
Devil's bit scabious	*Succisa pratensis*		
Didiscus	*Didiscus caeruleus*	Blue lace flower	
Dill	*Anethum graveolens*		Look out for the variety 'Florist's Dill'.
Dogwood	*Cornus* spp.		Grow bright cultivars for winter colour.
Echinacea	*Echinacea purpurea*	Purple cone flower	
Elder	*Sambucus nigra*	Common elder	Difficult to condition.
Eucalyptus	*Eucalyptus* spp.	Gum	
Euonymus	*Euonymus japonicus*		This is the variegated euonymus.
Euphorbia	*Euphorbia* spp.	Spurge	Some wonderful colours.
Everlasting flower	*Helichrysum bracteatum*		
Fennel	*Foeniculum vulgare*		
Feverfew	*Tanacetum parthenium*		You'll never curse it again.
Field marigold	*Calendula arvensis*		
Field poppy	*Papaver rhoeas*		
Forget-me-not	*Myosotis* spp.		
Forsythia	*Forsythia* spp.		
Foxglove	*Digitalis* spp.		
Freesia	*Freesia* spp.		
Garlic chives	*Allium tuberosum*	Chinese chives	
Geranium	*Geranium* spp.	Cranesbill	
Geum	*Geum* spp.		
Giant scabious	*Cephalaria gigantea*	Yellow scabious	Space-hungry darling.
Gladiolus	*Gladiolus* spp.	Sword lily	
Globe thistle	*Echinops ritro*	Blue hedgehog	
Grape hyacinth	*Muscari armeniacum*		
Guelder rose	*Viburnum opulus*		
Gypsophila	*Gypsophila elegans*	Annual baby's breath	We grow the 'Covent Garden' variety.
Harebell	*Campanula rotundifolia*		
Hawthorn	*Crataegus monogyna*	May; Beltane tree	
Hazel	*Corylus avellana*		
Helenium	*Helenium* spp.	Sneezeweed	
Hellebore	*Helleborus* spp.		
Holly	*Ilex aquifolium*		
Honesty	*Lunaria annua*		
Honeysuckle	*Lonicera* spp.		
Hornbeam	*Carpinus betulus*		

Commonly known as	Latin name	Also known as	Notes
Hosta	*Hosta* spp.		
Hyacinth	*Hyacinthus orientalis*		
Hydrangea	*Hydrangea* spp.		
Hyssop	*Hyssopus officinalis*		
Iris	*Iris* spp.		
Ivy	*Fatshedera lizei*	Tree ivy	
Kingcup	*Caltha palustris*	Marsh marigold	
Knapweed	*Centaurea nigra*	Common knapweed; hardheads	
Larkspur	*Consolida* spp.		Good also for confetti.
Laurel	*Prunus laurocerasus*	Cherry laurel	
Lavender	*Lavandula* spp.		
Lemon verbena	*Aloysia citrodora*		
Lilac	*Syringa vulgaris*		
Lily of the valley	*Convallaria majalis*	May lily	
Loosestrife 'Firecracker'	*Lysimachia ciliata* 'Firecracker'		We grow this for foliage, not flowers.
Mahonia	*Mahonia* spp.		
Mallow	*Malva* spp.		
Malope	*Malope trifida*	Annual mallow; mallow wort	
Marjoram	*Origanum vulgare*	Oregano	
Meadowsweet	*Filipendula ulmaria*	Bittersweet	
Melissa	*Melissa officinalis*	Lemon balm	
Mint	*Mentha* spp.		
Mistletoe	*Viscum album*		A great cash crop in winter.
Monkshood	*Aconitum napellus*		
Moon carrot	*Seseli libanotis*		
Myrtle	*Myrtus communis*		
Narcissus	*Narcissus* spp.		
Nasturtium	*Tropaeolum majus*		
Nerine lily	*Nerine bowdenii*	Bowden Cornish lily	
Nicotiana	*Nicotiana* spp.	Tobacco plant	
Nigella	*Nigella damascena*	Love-in-a-mist	Use the seedheads too.
Oak	*Quercus* spp.		
Old man's beard	*Clematis vitalba*	Traveller's joy	
Oriental poppy	*Papaver orientale*		
Orlaya	*Orlaya grandiflora*	White laceflower	

Commonly known as	Latin name	Also known as	Notes
Ornithogalum	*Ornithogalum* spp.		
Ox-eye daisy	*Leucanthemum vulgare*	Marguerite	This is the wild field daisy.
Paperwhite narcissus	*Narcissus papyraceus*		
Parsley	*Petroselinum crispum*		
Pear	*Pyrus* spp.		
Pendulous sedge	*Carex pendula*		
Penstemon	*Penstemon* spp.		
Peony	*Paeonia* spp.		
Persicaria	*Persicaria* spp.		
Phacelia	*Phacelia tanacetifolia*	Fiddleneck	Also good for green manure.
Philadelphus	*Philadelphus* spp.	Mock orange	
Phlox	*Phlox drummondii*		
Photinia	*Photinia* x *fraseri*	Christmas berry	
Physocarpus	*Physocarpus opulifolius*	Ninebark	
Pine	*Pinus* spp.		
Pink	*Dianthus* spp.		
Pittosporum	*Pittosporum tenuifolium*		
Pot marigold	*Calendula officinalis*	Calendula	'Art Shades' has interesting colours.
Purple loosestrife	*Lythrum salicaria*	Black blood	
Quaking grass	*Briza media*	Pearl grass	
Ragged robin	*Lychnis flos-cuculi*	Crow flower	
Ranunculus	*Ranunculus asiaticus*	Persian buttercup	
Red campion	*Silene dioica* (syn. *Melandrium rubrum*)	Adder's flower	
Red valerian	*Centranthus ruber*		
Rose	*Rosa* spp.		
Rosebay willowherb	*Chamaenerion angustifolium*		
Rosemary	*Rosemarinus officinalis*		
Rowan	*Sorbus* spp.	Mountain ash	
Rudbeckia	*Rudbeckia hirta*	Black-eyed Susan	
Sage	*Salvia* spp.		
Scabious	*Scabiosa* spp.		
Schizostylis	*Hesperantha coccinea*	Crimson flag lily	
Scilla	*Scilla* spp.		
Sea holly	*Eryngium* spp.		
Sedum	*Sedum* spp.		
Skimmia	*Skimmia japonica*		

Commonly known as	Latin name	Also known as	Notes
Smoke tree	*Cotinus coggygria*		Difficult to condition.
Snake's head fritillary	*Fritillaria meleagris*	Chequered daffodil	
Snapdragon	*Antirrhinum* spp.		
Snowball bush	*Viburnum opulus* 'Roseum'		Watch out for viburnum beetle.
Snowdrop	*Galanthus nivalis*		
Snowflake	*Leucojum aestivum*		
Solomon's seal	*Polygonatum* x *hybridum*		
Sorrel	*Rumex acetosa*		
Spindle	*Euonymus europaeus*		
Spiraea	*Spiraea japonica*	Japanese spiraea	
Spotted laurel	*Aucuba japonica*		
Statice	*Limonium* spp.		
Strawberry	*Fragaria* x *ananassa*		
Sun spurge	*Euphorbia helioscopia*		
Sunflower	*Helianthus annus*		
Sweet pea	*Lathyrus odoratus*		
Sweet rocket	*Hesperis matronalis*	Dame's violet	
Sweet William	*Dianthus barbatus*		
Teasel	*Dipsacus fullonum*		
Tellima	*Tellima grandiflora*	Fringe cups	
Thistle	*Cirsium vulgare*	Spear thistle	
Thrift	*Armeria maritima*	Sea pink	
Thyme	*Thymus* spp.		
Tulip	*Tulipa* spp.		
Valerian	*Centranthus ruber*		
Verbascum	*Verbascum* spp.	Mullein	
Verbena bonariensis	*Verbena bonariensis*	Argentinian vervain; purple top	
Veronica	*Veronica gentianoides*	Gentian speedwell	
Viburnum bodnantense	*Viburnum* x *bodnantense*	Arrowwood	
Wallflower	*Erysimum* spp.		
White campion	*Silene latifolia*	Bull rattle	
Wild carrot	*Daucus carota*	Queen Anne's lace	
Willow	*Salix* spp.		
Winter-flowering honeysuckle	*Lonicera fragrantissima*		
Zinnia	*Zinnia* spp.		

Resources

We farm flowers in Somerset, England, and so the resources I give here are largely UK-based. There are, great resources in other Anglophone countries for artisan flower farmers. Readers in the US, Australia and New Zealand should visit www.greenbooks.co.uk/flower-farmers-us-anz-resources for more information.

I also urge you to research your own local growers and plant nurseries. There is a wealth of advice out there, and gardeners are people who simply cannot resist sharing their knowledge, which makes them some of the best people to count as colleagues. Don't be afraid to ask other growers where they go to buy stock!

Seeds

Avoid buying seed from garden centres, as it may have been sitting there in a very hot, bright place for a long time – not the optimum conditions for seed storage. The following are a few of my favourite suppliers.

Chiltern Seeds
www.chilternseeds.co.uk   Tel. +44 (0)1491 824675
For the best-quality seed. Recommended for the grower who really wants to curate their garden to the highest standard with the most interesting plants.

Emorsgate Seeds
http://wildseed.co.uk   Tel. +44 (0)1553 829028
They source their seed locally to specific areas, so you can make sure that your wildflower seed is not only native to the UK but also to your immediate locality.

Higgledy Garden
www.higgledygarden.com
For fun, for beginners, and for a giggle-inducing website. Benjamin Ranyard's seeds are great value because he doesn't put more than you need into each packet.

Kings Seeds
www.kingsseeds.com   Tel. +44 (0)1376 570000
For good-quality wholesale seeds. They offer a wide variety of sweet pea seed, as well as organic seed.

Moles Seeds
www.molesseeds.co.uk   Tel. +44 (0)1206 213213
Also for good-quality wholesale seeds.

Tamar Organics
www.tamarorganics.co.uk   Tel. +44 (0)1579 371098
A good choice of organic flower seed (not just veg).

Roses

The following two suppliers are the most well-known names for roses in the UK. They will give you good advice and both pre- and post-sale support. Do look around locally as well, though: you may find you have a nearby nursery growing roses that will do well in your local conditions, and from whom you will get really good advice on an ongoing basis. Just because a supplier is the most famous doesn't mean they will always have the best product for you.

David Austin Roses
www.davidaustinroses.co.uk
Tel. +44 (0)1902 376301

Peter Beales
www.classicroses.co.uk   Tel. +44 (0)1953 454707

Dahlias

Both the following supply good-quality rooted dahlia cuttings, which they deliver in April or May. Remember to order your dahlias well before Christmas, because your suppliers may run out of certain varieties that are popular that year.

The National Dahlia Collection
www.national-dahlia-collection.co.uk
Tel. +44 (0)7879 337714

Withypitts Dahlias
www.withypitts-dahlias.co.uk
Tel. +44 (0)1342 714106

Plants grown in the UK

Habitat Aid
www.habitataid.co.uk   Tel. +44(0)1749 812355
An online resource for a great many plants, all guaranteed to be grown in the UK. For UK growers who care about the provenance of the plant material they buy in, Habitat Aid is a good place to start.

Municipal green-waste compost

The following two national suppliers are good starting points. Check also the back of your local advertiser magazine for local suppliers of spent mushroom compost – or ring your local mushroom farm.

Melcourt
www.melcourt.co.uk   Tel. +44 (0)1666 502711
A wide range of products: check their website for local suppliers.

Viridor UK
www.viridor.co.uk   Tel. +44 (0)1823 721400
Recycling specialists – they don't deliver compost nationwide, but it's worth ringing them and asking.

Florists' sundries

It's worth setting yourself up with a small stock of buckets, scissors, raffia, ribbon, cellophane, etc., as part of your start-up costs. There are lots of different florists' sundries suppliers online, all of which do a reasonable job, and none of which I'd recommend over another. Shop around for the best prices.

Other supplies

Wiggly Wigglers
www.wigglywigglers.co.uk   Tel. +44 (0)1981 500391
For ladybird larvae – and other natural pest-control products, plus much more.

Twool
www.twool.co.uk   Tel. +44 (0)1364 654467
For grown-in-the-UK wool garden twine.

Help and volunteers

The Women's Farm and Garden Association (WFGA)
www.wfga.org.uk   Tel. +44 (0)1285 658339
Men welcome too! A long-established organization giving gardeners training on the job. A great resource if you're looking to train yourself, or to find people whom you're happy to train while they work.

Work Exchange on Organic and Sustainable Properties (WWOOF)
http://wwoofinternational.org
Wwoofers are volunteers on organic farms who will work in return for bed and board. People may come for as little as a fortnight, or for up to several months, to help for several hours a day. Wwoof is a network of national organizations: the website will direct you to the local website for your area.

Flower-growing communities

You might join one of various collectives, such as the following (the first two in the UK, the third in the US).

The British Flower Collective
www.thebritishflowercollective.com

Flowers from the Farm
www.flowersfromthefarm.co.uk
info@flowersfromthefarm.co.uk

Association of Specialty Cut Flower Growers
www.ascfg.org +1 440 774 2887

Be wary of spending too much money with join-up fees for an organization until you're sure that you like growing flowers as a crop, and want to spread the word about what you do. In the short term I recommend using Twitter, Facebook, Pinterest and Instagram to tell the world what you're up to on your cut-flower patch – and also to find fellow growers and supportive growing communities.

On Twitter in the UK join a regular chat between flower growers on Monday nights between 8 p.m. and 9 p.m., using the #tag #britishflowers. We have American and Dutch growers who join in too and are very welcome.

Blogs to be inspired by

www.commonfarmflowers.com/blog.php
My own blog. An ongoing diary of all the goings-on at Common Farm, from planting to floristry. For the story of a flower farm from its beginnings you may find it useful.

www.floretflowers.com/blog
Erin Benzakein's blog from her flower farm near Seattle, WA: beautiful, inspiring and informative.

www.littleflowerschoolbrooklyn.com/journal
The teaching project of two sister Brooklyn-based businesses, who teach, create wonderful floristry, and inspire in equal measure.

http://thephysicblog.blogspot.co.uk
Sara Venn's blog – for gardening tips and information. Sara is a horticulturalist and consultant working and teaching in Bristol. I go to her for horticultural advice.

Colleagues

Lastly, and gratefully – the kind contributors to this book:

Easton walled gardens

www.eastonwalledgardens.co.uk
Tel. +44 (0)1476 530063
Ursula Cholmeley's creation. For a wonderful display of flowering sweet peas throughout the summer – and very good scones in the tea room too.

Flowers by Clowance

www.flowersbyclowance.co.uk
Tel. +44 (0)1209 831317
Wholesale British-grown flowers.

Peter Nyssen Flower Bulbs & Plants

www.peternyssen.com   Tel. +44 (0)161 747 4000
Wholesale bulb supplier to private customers and the trade.

Reads Nursery

www.readsnursery.co.uk   Tel. +44 (0)1986 895555
Specialist fruit-tree supplier.

Withypitts Dahlias

(see page 248)
Good-quality dahlia plants.

Books

The following are a few of my favourite books on horticulture and floristry.

Gilding the Lily: Inside the cut flower industry
Amy Stewart
Portobello Books (2009)
A fascinating insight into the global cut-flower industry.

On growing

The Flower Farmer: An organic grower's guide to raising and selling cut flowers
Lynn Byczynski
Chelsea Green Publishing (1997, second edn 2008)
A really excellent introduction to flower farming, by a grower in Texas.

Grow Your Own Cut Flowers
Sarah Raven
BBC Books (2002)
A classic and a great resource.

Specialty Cut Flowers: The production of annuals, perennials, bulbs and woody plants for fresh and dried cut flowers
Allan M. Armitage and Judy M. Laushman
Timber Press (1993, revised edn 2008)
A really good book-long list of flowers to grow, with details of how to cut and condition, vase-life, etc. Highly recommended.

Sweet Peas: An essential guide
Roger Parsons
The Crowood Press (2011)
An excellent book on sweet pea growing, harvesting, disease management, etc.

Woody Cut Stems for Growers and Florists: Production and post-harvest handling of branches for flowers, fruit, and foliage
Lane Greer and John M. Dole
Timber Press (2008)
Fascinating, all-encompassing and inspiring, for people with all kinds of land who want to grow shrubs and foliage for cutting.

Floristry

Look for books by Vic Brotherson of Scarlet and Violet in London, published by Kyle Books. Her blithe, loose style of floristry is perfect for the home-grown, fresh-from-the-garden look.

Vintage Flowers: Choosing, arranging, displaying
Kyle Books (2011)

Vintage Wedding Flowers: Bouquets, buttonholes, table settings.
Kyle Books (2014)

Also look in second-hand and vintage shops for books by Constance Spry: the godmother – or patron saint perhaps – of truly inventive floristry. Her books are breathtaking examples of how less is often so much more, how rules are irrelevant, and how with flair, a good eye and an assumption that your idea will work, you can create wonders with almost any cut stem from the garden.

Index

Page numbers in **bold** indicate the main source of information on a recommended plant. Page numbers in *italic* refer to illustrations.

accounts 206, 209-10, 218
achillea *74*, **74**, 146
acidanthera 86, 94, *95*, **95**
acorns 171-2
agapanthus *95*, **95**
  'Apple Blossom' 96
alchemilla 22, 43, 69, 73, **74-5**, 117, *153*, 154, *154*
alliums 86, 87, **88**, 101
  A. cowanii 88
  A. sphaerocephalon 88, *89*
  garlic chives 88
amaranthus **51**
amaryllis 97, *98*, **98-9**, 101
ammi *10*, 16, 38, 39, 40, *43*, **43**, 55, 68, 137, *153*
  A. majus 18, 42, 43
  A. visnaga 43
anemones *33*, 87, *89*, **89**, *111*, 101
angelica **159**
annuals 24, 36-57
  growing from seed 38, 39-42, 47-8, 53-6, 57
  half-hardy annuals 39, 40, 47-51
  hardy annuals 39, 40, 41-7
  overwintering 42
  planting out 48
  quick-turnaround crop 37, 38, 68
  recommended annuals 42-7, 49-53
  tender annuals 39, 40, 48, 51-3
  top tips 57
aphids 35, 74, 125, 146, 163, 168
  control 167, 168, 169
apple mint 158, 161
aquilegia *74*, **74**, 178
artemisia **75**
artisan florists and flower farmers 10
asters 38, **49**, 69, 71, 138
  'Florette Champagne' 49
astrantia 73, **75**
  A. 'Hadspen Blood' 75
  A. major 75
Austin, David 117

bachelor's buttons *see* cornflowers
bank account 210
barn owls 167, 184

basil 163
bay **106**
beds
  construction 23
  no-dig beds 25
  raised beds *18*, 22, *22*, 57
  seedbeds 40-1
  sizes 16, 228
  space between 18
beech 171
bees 53, 137, *166*, 167, 168
bells of Ireland 39, **49**, 137
biennials 58-65
  growing from seed 61-2, 65
  planting out 61
  recommended biennials 63-5
  as short-lived perennials 62, 63, 65
  top tips 65
  vegetative growth 61
bistort 79, 178, 181
black spot 117, 118
blackberry **172**
blackthorn 173
bleeding heart **75**
blogs 232, 249-50
bluebells *174*, **174-5**
botrytis 34, 40, 56, 91, 146
brachyglottis 20
bridal market 121, 190-1, 226-7
brodiaea **89**
bronze fennel 137, **159**, *160*
buckets *184*, 186-7, 195
buddleja **106**
buddleja mint 161
budgeting 211-12
bulbs and corms 38, 82-101
  as annuals 84, 87, 101
  buying 85-6
  Christmas market 97-100, 199
  financial return 84, 85
  forcing 98, 99, 101
  growing 85-8, 100, 101
  leaving in the ground 85
  naturalizing 86-7, 100, 101
  recommended bulbs 88-97, 98-100
  spring-flowering 86, 88-94
  summer- and autumn-flowering 86, 94-7
  top tips 100
  underplanting other crops 94, 100
bupleurum **43**
business expenses 210

business plan 209, 218
buttercups *175*, **175**, 178, 181
butterflies *166*, 168
buying plants 71-2
  bulbs 85-6
  plant sales 69, 71, 96
  wholesale prices 71

calendula 42
California poppies **63**
  'Aurantiaca' 63
  'Champagne and Roses' 63
  'Rose Chiffon' 63
camassia **89-90**
campanula **75**
Canterbury bells *see* campanula
catkins 113, 173, 174
catmint **160**
cellophane 194
cerinthe 18, 42, *43*, 154
chalky soil 22, 181
Chelsea Chop 70, 81
chicken wire 147
chilli and garlic dip 21, 86
choisya 20, **106**
Cholmeley, Ursula 155
Christmas box 105, **106**, 107, 199
Christmas decorations 52, 73, 79, 96, 101, 109, 180, 196-203
  bulbs 97-100, 199
  toolkit 200-1
  top tips 203
  see also wreaths and garlands
chrysanthemums **44**, *123*
  'Polar Star' 44
  'Primrose Gem' 44
  see also feverfew
clary sage 38, *44*, **44**, 162
clay soil 22, 25, 181
clematis 112, 199
  old man's beard 171, *172*, 199, 203
cleome 18, *49*, **49-50**
Cock, James 219
coir compost 56
cold frames 30
columbine *see* aquilegia
comfrey 23, 27, **160**, *170*
comfrey tea 27, 132, 146
  recipe 28
Common Farm Flowers 11-12, *16*, *17*, 166-7
  planting list 239-41

social media story 230-1
compost 22, 25, 26
    coir compost 56
    compost heap 26-7, 168, *170*
    green-waste compost 23, 26, 56, 167, 249
    home-made 27, 56
    mushroom compost 26
    peat-free 56
    seed compost 40, 56, 61
    sterilization 56
compost tea 23, 26, 27, *28*, 132, 146
    recipe 28
conditioning flowers 188-90, 195
    flower-preserving chemicals 188, 189-90
    searing stems 125, 189
    see also individual flowers
confetti 44, 45, *122*
copyright 233
coriander **160**
corkscrew hazel **106**
corkscrew willow **106**
cornflowers 16, 18, 39, 40, 42, **44-5**
    'Black Ball' 44, *45*
cosmos 16, 38, 39, 40, 41, *41*, **52**, 68, *138*
cow parlsey 43, **75**, *174*, *175*, **176**
    'Ravenswing' 75, 176
cowslips *93*, *174*, 176, **176**
crab apples 113, 122, 171, 199
craspedia **50**
crocosmia 94, **96**
crocuses 21, 97
crop protection *see* cold frames;
    greenhouses; polytunnels;
    windbreaks
cropping schedule 48, 236-9
cut-flower business 204-19
    accounts 206, 209-10, 218
    bank account 210
    budgeting 211-12
    business expenses 210
    business plan 209, 218
    employees 215-16, 217
    environmental health 217
    insurance 217
    intellectual copyright 233
    international cut-flower industry 9, 10, 12
    investment 210-11, 218
    legal issues 25, 97
    marketing *see* marketing
    packaging and delivery 218
    planning permission 217
    pricing 213-15, 222
    pros and cons 208
    realistic expectations 207-8

savings account 212
taxation 213
top tips 218, 219
VAT 213
cutting flowers 185-8
    cutting list 188
    directly into water 185, 195
    for weddings and other events 190-1
    stem length 188
    time of day 185, 195
    toolkit 185-7
    top tips 195
    woody stems *124*, 125, 185, 188
    see also individual flowers

daffodils and narcissi 84, 86, 88, **90**, *93*, 97, 101
    'Early Cheerfulness' 90
    'Grand Soleil D'Or' 100, 101
    paperwhite narcissi 97, **100**, 101
    'Ziva' 100
dahlias 13, 16, 22, 24, 60, 62, 68, 94, **126-39**, 154, 214, 248
    'Apricot Desire' 136, *136*
    'Barbarry Olympic' 136
    'Blackberry Ripple' 136
    bouquet combinations 137-8
    cactus dahlias 134-5, *134*, 139
    'Carolina Wagemans' 139
    collarette dahlias 134
    cutting and conditioning 136-7, 139
    cuttings 128
    decorative dahlias 135
    dinner plate dahlias 135
    disbudding 130, *131*
    dwarf dahlias 136
    'Glorie Van Heemstede' 139
    growing 128-33, 138
    Karma series 135, 139
    lifting and storing tubers 128, 132-3, *132*, 138
    'Orfeo' 136
    planting out 128, 129-30, 138
    pompom, ball and miniature ball dahlias 133-4, 139
    propagation 129
    recommended dahlias 136
    semi-cactus dahlias 135
    staking 130, *131*, 132, 138
    'Summer Night' ('Nuit d'ete') 136
    'Taratahi Ruby' 139
    top tips 138
    waterlily dahlias 135, 139
damping off 56
deer 20-1
delphiniums *24*, 69, *69*, 71, **75-6**
    see also larkspur

didiscus **50**
dill 159, **160**
dogwood *104*, 111, 171, **172**

earwigs 137
echinacea 69, *70*, 71, 73, **76**, 138, *166*
elder **106-7**, **172**
employees 215-16, 217
environmental health 217
environmental impact of the cut-flower industry 9
ephemeral beauty of cut flowers 10-11
euonymus 20, **107**, 199, 203
euphorbia 76, **76**, 167
    E. dulcis 'Chameleon' 76
    E. marginata 76
    E. oblongata 76
everlasting flowers *38*, *52*, **52**, *138*

Facebook 230, 231
farmers' markets 45, 149, 150, 191, 198, 224
fashion 65, 85, 121, 135, 138
fencing 20-1, *21*
feverfew *10*, **160**
first aid 217
florists, local 222-4
flower foam 192-3, 194
'flower food' 188, 189-90
flower frogs 193
flower-growing collectives 249
foliage 19, 63, 70, 75, 78
    see also herbs; shrubs
forcing 33, 91, 98, 99, 101, 112
forget-me-nots *174*, **176-7**, *177*
forsythia 112
foxgloves 41, 61, **63-4**
    'Pam's Choice' *64*
freesias **90**
fritillaries *92*, **92**, *174*, **179-80**
frost 39, 40, 48
fruit blossom 112, 113, 122
fungicide 9, 10

garlic chives 88
gate sales 45, 150, 224-5
gel pearls 195
geraniums **76**
German pins 193
geums *77*, **77**
gladioli 86, 88, 94, *96*, **96-7**, 101
    Gladiolus nanus 96
    Nanus varieties 101
granny's bonnet *see* aquilegia
grape hyacinths 85-6, 101
grass snakes 27, 167, 169-70, *169*
green-waste compost 23, 26, 56, 167, 249

greenhouses 30-1, 32
  hygiene 34
  pros and cons 32
  ventilation 35
  see also polytunnels
guelder rose 20, *104*, 171, **173**
gypsophila 39, 40

hardiness zones (US) 13
harebells **177**, 181
hawthorn 111, 171, **173**
hazel **106**, 167, **173**
hazel prunings 24, 148, *148*
health and safety 217
hedgehogs 27, 167, 169
hedging 19, *19*, 20, 104, 166, *170*
  roses 122
  wild hedge plants 171-4
heleniums 70, **78**
hellebores **77**
herbs 16, 22, 117, **156-63**
  cutting and conditioning 163
  growing 158, 163
  overwintering 163
  recommended herbs 159-63
  top tips 163
  useful properties 163
holly 171, 172, 199
honesty **64**, 199, 203
honesty boxes 225
honeysuckle *107*, **107**
honeywort *see* cerinthe
hornbeam **107-8**
hostas **78**
hyacinths 33, **90**, 97, *99*, **99**, 101
hydrangea 80, **108**, 199, 200, 203
hygiene 34, 55, 133, 185, 186, 195
hyssop **160**

industry press 234
Instagram 230, 231
insurance 217
international cut-flower industry 9, 10,
  12
irises **90**
  bearded irises (*Iris germanica*) 90
  Dutch irises (*Iris* x *hollandica*) 90
  wild (foetid) irises (*Iris foetidissima*)
    90, 181
ivy 167, *168*, 199, 203

kestrels 167
kingcups 181
knapweed **177**

labels, plastic 56
ladybirds 35, 76, 146, 167, 168, 169

lady's mantle *see* alchemilla
larkspur 18, 39, 42, **45**, 68
Latin names 13, 242-7
laurel 105
lavender 20, 158, 159, **160-1**, 163
  'Grosso' 160
leaflets 233-4
legal issues 25, 97
lemon verbena **161**
Leucanthemum x *superbum* 'Osiris
  Neige' 79
lilac *108*, **108**, 111, *111*, 112, 178, 189
lily beetles 179-80
lily of the valley **91**
liquid feeds *see* comfrey tea; compost
  tea; nettle tea; seaweed solution
local press 230, 232-3
loosestrife **78**
  'Firecracker' 70, 78, *78*
  purple loosestrife **178**
love-in-a-mist *see* nigella
love-lies-bleeding *see* amaranthus
Lynes, Karen 101

mahonia **108**
mallows **50**, *70*
Malope trifida 48
manures 22, 23, 25, 26, 97
  pathogens 25, 97
marjoram 158, 162
marketing 206, 218, 220-34
  Christmas 198
  customer base 214, 221-7
  farmers' markets 45, 149, 150, 191,
    198, 224
  florists, local 222-4
  gate sales 45, 150, 224-5
  industry press 234
  leaflets 233-4
  local press 230, 232-3
  networking groups 234
  pick-your-own 225-6
  social media 229-32, 234
  top tips 227, 234
  websites 232
  wedding fairs 226-7
May flower *see* hawthorn
meadows 170-1, 181
meadowsweet **177**
melissa 158, **161**
mice 35, 47, 85, 100, 133, 143, 167, 169
microclimates 12
mildew 91, 147
mint 117, 158, 159, **161**, 163
  apple mint 158, 161
  buddleja mint 161
  pineapple mint 158, *158*, 161

mistletoe 199
mock orange *see* philadelphus
monkshood 79
moon carrot 153
moss 193
mould 34, 56, 143, 163
mulching 23, 24, 25, 26, 34, 41, 118
mushroom compost 26
mycorrhizal fungi 26, 118, 119
myrtle **109**

narcissi *see* daffodils and narcissi
nasturtiums *50*, **50-1**
nerine lilies 94, **97**
nettle tea 27, 132, 146
  recipe 28
nettles 27, 168
networking groups 234
nicotianas 29, **52**
nigella **45**, 54, *54*, 104, 199
  'Miss Jekyll' 45
no-dig gardening 25
non-UK growers 12-13

oak 171
old man's beard 171, *172*, 199, 203
oregano **162**
organic farming 71
oriental poppies **78-9**, *79*
orlaya 18, **46**, 153
ornithogalum **97**
ox-eye daisies **79**, *175*, **177-8**

packaging and delivery 218
paperwhite narcissi 97, **100**, 101
parsley 158, *158*, **162**
pea netting 23, 24, *24*, 130, *131*, 147
peat-free compost 56
pens, indelible 56
penstemons **79**, 81
peonies 73, 93, 154, 214
perennials 38, 66-81
  buying plants 69, 71-2
  Chelsea Chop 70, 81
  cropping 72, 73
  cuttings 69
  feeding 81
  growing from seed 69, 71, 81
  herbaceous borders 69
  planting 68-9
  propagation 72
  recommended perennials 73-81
  splitting 69, 72
  top tips 81
persicarias **79-80**
  bistort 79, 178, 181
  'JS Caliente' 79

pests
  biological pest control 35, 167, 168-9
  deterrence 20-1, 86, 163
  see also individual index entries
phacelia 42, *46*, **46**
philadelphus **109**
phlox **80**
photinia 104
photography 227
physocarpus 20, *20*, **109**, *138*, *175*
  'Dart's Gold' 109
pick-your-own 225-6
pinching out 39, 144-5, *145*
pineapple mint 158, *158*, 161
Pinterest 230, 231
pittosporum 20, **109**
  'Tom Thumb' 109
planning permission 217
plant breeders' rights (PBR) 25
plant sales 69, 71, 96
planting rotation 21-2, 33, 94
pliers 200
plot design and practicalities 16-24
  choosing what to grow 18-19
  example plot layout *35*
  fencing 20-1, *21*
  planting rotation 21-2, 33, 94
  plot size 11, 16
  staking plants 23-4
  windbreaks 19-20
  see also beds; compost; soil
poisonous plants 79
polytunnels *17*, 30, *31*, *33*, 169, *210*
  heating 34
  hygiene 34
  pests and diseases 34-5
  pros and cons 31
  using 32
  ventilation 33-4, 35
  watering system 34
ponds 168, *170*
poppies 189, 199
  see also California poppies; oriental
    poppies
post, sending flowers by 218
pot marigolds 16, 39, *46*, **46**
  'Sherbet Fizz' 46
presenting cut flowers 191-5, 206
  binding 194
  finishing 194-5
  flower foam 192-3, 194
  flower frogs 193
  packaging and delivery 218
  vases and containers 191, *191*, 194
  wire and supports 192-3
pricing 213-15
  per stem 214-15, 222

propagating beds 29, *29*, 30, 48
propagating mats 29
propagation 29, 30-1
  dahlias 129
  perennials 72
  see also seed, growing from

quaking grass **46-7**

rabbits 20-1
raffia 194, 218
ragged robin *178*, **178**
Ramsey, Richard 137, 139
ranunculus 87, *91*, **91**, 101
rats 133
Read, Stephen 113
red campion *174*, **178**
red valerian **178**
refrigeration units 98, 99
resources 248-50
ribbons 194, 200, 218
Rootrainer 143
rosebay willowherb **80**, **178-9**
rosemary 158, **162**, 163, 199
  'Miss Jessopp's Upright' 20, 162
roses *10*, 16, 71, 73, 93, *111*, **114-25**,
    154, 214, 248
  'A Shropshire Lad' *116*, 122, **123**
  black spot 117, 118
  'Bonica' **123**
  bridal market 121
  'Buff Beauty' *120*
  choosing which to grow 120-1
  'City of York' *121*, **123**
  climbers and ramblers 122
  'Compassion' 119, 122, *123*, **123**
  cutting and conditioning 124-5, *124*
  growing 117-19, 125
  hedging 122
  hips 121, 122, 199, 200, *200*
  'Madame Alfred Carrière' *121*, *154*
  'Munstead Wood' **123**
  old rose varieties 121
  pruning 119
  'Rambling Rector' 122
  recommended roses 122-3
  Rosa mundi 122
  rose petal confetti *122*
  spraying 125
  thorn removal 125
  top tips 125
  underplanting 22, 117, 125, 163
  'Wild Edric' 118
  wild roses 171
rowan 171
rudbeckias **51**
  'Cherry Brandy' 51

sage 117, 158, **162**, 163
  clary sage 38, *44*, **44**, 162
  'Hot Lips' 162
  'Victoriana' 162
scabious *10*, 18, **47**, **80**, *138*, 179, 199
  devil's bit scabious *170*
  'Drumstick' *47*, 50
  field scabious *179*
  giant scabious **77**
  'Pink Mist' *68*
  'Snow Maiden' *47*
schizostylis 94, **97**
scissors 185-6, 195
sea hollies 71, 73, *73*, **80**, 138
searing stems 125, 189
seaweed solution 146
secateurs 185, 186, 200
sedums 73, **80**, 138
seed
  buying 53-4, 55, 248
  harvesting 54
  saving 53, 54-5, 57
  stale seed 55
seed compost 40, 56, 61
seed, growing from 38
  autumn sowing 40-1, 42
  direct sowing 39-41, 42
  equipment 55-6
  pre-treating seed 142-3
  seed drills 41
  staggered sowing 39, 42
  under cover 9, 39, 40, 41-2
  vernalization 45
  watering 56
  winter/spring sowing 41, 42
  see also under annuals; biennials;
    perennials; sweet peas
seed trays 34, 55-6, 143
seedbeds 40-1
seedheads 50, 51, 64, 79, 80, 96, 101,
  109, 180, 199, 203
seedlings
  potting on 40, 56
  pricking out 56
  protection 22
  self-sown 42
  thinning 41
  watering 40
shade mesh 22
shrubs 38, 102-13
  cutting and conditioning shrubby
    material 112
  evergreens 105
  forcing 112
  fruit blossom 112, 113, 122
  growing 104-5
  hedging shrubs 20

recommended shrubs 105-11
    top tips 112
    variegated 105, 107, 109
    windbreaks 104, 105
silver nitrate 10, 47, 137, 152, 190
skimmia **109**
skin reactions 74, 76
sloes 173
slow-worms 27, 167
slug traps 57
slugs 27, 40, 57, 167
    deterrence 57, 78, 86
    eggs 56
smoke tree **110**
snake's head fritillary *92*, **92**, **179-80**
snapdragons 38, **47**
snowball bush *see Viburnum opulus*
    'Roseum'
snowdrops *92*, **92**
snowflakes **92**, 101
soap spray 169
social media 229-32, 234
soil
    conditioning and management 22,
        23, 25, 26, 41
    drainage 22, 41, 100
    pH 24, 26
    topsoil 23, 26
    see also liquid feeds; mulching
soil testing kits 26
soldier bugs *168*
Solomon's seal 75, **80-1**
sorrel **180**
spindle 171, 199
spiraea **110**
    S. 'Bridal Wreath' *93*, 110
    S. *douglasii* 110
    S. 'Goldflame' *110*
    S. *japonica* 'Candlelight' 18
squirrels 85, 90
staking plants 23-4, 130, 132
    chicken wire 147
    individual staking 24
    pea netting 23, *24*, 130, *131*, 147
    string/twine 23, 132
    teepees 24, 130, 148, 154
statice *39*, **51**, 199, 203
    'Sunset' 51
Stemtex 194
strawberry flowers 77, 163, *163*
'strulch' 24
sugar solution 152, 190
sun spurge *76*, 178
sunflowers 16, *53*, **53**
    pollen-free varieties 53
superstition 92, 108, 173
sweet peas 11, 16, 20, 22, 38, 42, **47**, 68,

93, *123*, **140-55**, 214, 215
'Beaujolais' 68, 151, *151*
'Betty Maiden' 151-2, *154*
bouquet combinations 153-4
'Charlie's Angel' *150*, 152
choosing which to grow 149-52, 154
colours 150-1
cordon-grown 145
'Cupani' 149
cutting and conditioning 152
'Daily Mail' 68
feeding 146
grandiflora varieties 149, 154, 155
growing 142-9, 154, 155
'Gwendoline' 155
heritage varieties 149, 150, 154, 155
'King Edward VII' 155
'King's High Scent' *154*
'Linda Carole' 155
'Lord Nelson' 155
'Matucana' 155
'Mollie Rilstone' 142, *142*, 152, 155
'Our Harry' 155
'Painted Lady' 149, 155
pinching out 144-5, *145*
planting 145-6, 154
recommended sweet peas 151-2
'Royal Wedding' *150*, 152
saving seed 148
sowing seed 39, 47, 142-3, 144, 155
Spencer varieties 142, 149-50, 154
successional crops 144, 148
support structure 146-8, *148*, 154
top tips 154, 155
sweet rocket **64**, *153*
sweet Williams 41, 60, *60*, 62, **64-5**
'Auricula-Eyed Mixed' *60*, 65
'Green Trick' 65
'Sooty' *62*, 65

taxation 213
teasel **180**, 199
teepees 24, 130, 148, 154
tellima *78*
test-tube containers 194
thistle 73, 138, **180**
thrift **65**
thyme 158, **163**
tissue paper 194
tithe to nature 53, 166-70, 181
toads 167, 169
trolleys 187, *187*
tulip fire 93-4
tulips 84, *84*, 86, 87, 88, *93*, **93-4**, 101,
    117
    Darwin Hybrids 101
    'Spring Green' *94*

twine 23, 132, 194
Twitter 230, 231, 234, 249

Vaseline 57
vases and containers 191, *191*, 194
    see also buckets
VAT 213
verbascum **81**
    'Purple Temptress' 81
verbena bonariensis *43*, **51**, *81*, **81**,
    137-8, *138*
vermiculite 56, 61
vernalization 45
veronicas **81**
viburnum 20, *104*, 107, **110-11**
    guelder rose 20, *104*, 171, **172**, 20,
        104, 171, **173**
    V. *bodnantense* 111
    V. *opulus* 'Roseum' (snowball bush)
        20, 110-11, *111*
viburnum beetle 111, 173
volunteers 217, 249

walled gardens 19
wallflowers **65**
websites 232
wedding fairs 226-7
weeds 60
    hand weeding 42
    hoeing 42
white campion 178
wild carrot 137, *168*, *172*, **180**, 199
wildflowers **164-81**
    cutting and conditioning 180-1
    hedge plants 171-4
    meadows 170-1, 181
    recommended wildflowers 174-80
    top tips 181
wildlife 26-7, 166-8, 181
    see also individual index entries
willow **106**, 112, 167, **173-4**, 199
willow wreaths *199*, 201-2, *202*
wind rock 20, 22, 23
windbreaks 19-20, 104, 105
window sills 48
wire 147, 192, 193, 200
Women's Farm and Garden Association
    (WFGA) 215, 249
wreaths and garlands *198*, 199, *199*,
    200, *200*, 201-3, *202*

year planner 13, 236-9
yellow rattle 171

zinnias 39, **53**
    Benary's Giant series 53

Also by Green Books

Grow Your Own Wedding Flowers
Georgie Newbery

This inspirational practical book will take you through everything you need to grow and arrange your own wedding flowers. *Grow Your Own Wedding Flowers* is filled with guidance and information on selecting, growing, cutting, conditioning and arranging your flowers, from buttonholes and bouquets to pew-ends and confetti. Seasonal planting plans cover spring, summer, autumn and winter weddings, allowing you to pick the perfect selection for your special day – and helping you save money.

Published February 2016

Gardening for Profit: from home plot to market garden
Kate Collyns

This book is a must-have for anyone who is interested in selling some produce for profit – whether just surplus from a vegetable garden or wholesale from a fully developed professional business. It takes you step by step through all the aspects you'll need to think about, including: finding land, winning customers and marketing your produce, working out what equipment you'll need, budgeting, sourcing funding, managing your accounts, doing the tax and legal paperwork, and deciding which crops to grow. It is an invaluable resource for anyone wanting to sell their produce.

Wild Flowers: an easy guide by habitat and colour
Tracy Dickinson

This full-colour guide to the wild flowers of Great Britain is designed to make identification as easy as possible. The flowers are listed in eight sections, according to the one or more habitats in which they are found. Each habitat section has a set of introductory photographs for easy identification, followed by a set of larger photos together with essential information on their characteristics and flowering months. Perfect for anyone who loves wild flowers.

About Green Books

Environmental publishers for 25 years.
For our full range of titles and to order direct from our website, see **www.greenbooks.co.uk**

Join our mailing list for new titles, special offers, reviews, author appearances and events:
www.greenbooks.co.uk/subscribe

For bulk orders (50+ copies) we offer discount terms. Contact **sales@greenbooks.co.uk** for details.

Send us a book proposal on eco-building, science, gardening, etc.: see **www.greenbooks.co.uk/for-authors**

 @ Green_Books   /GreenBooks